Exceptional mass timber volume is coming

As one of the oldest mass timber manufacturers in the U.S.,
we've produced 785,000 panels to date and our annual capacity
of cross-laminated timber is over 54 million sq. ft. per year.

SAVE THE DATE
FOR THE 8TH ANNUAL

— INTERNATIONAL —

MassTimber ℠
CONFERENCE

March 27–29, 2023
(Monday–Wednesday)
Portland, Oregon USA
Oregon Convention Center

MassTimberConference.com

Editing by Self-Publishing Services LLC

Formatting by Meld Media and Made Graphic Design

Cover Photo Credit: Ema Peter Photography, courtesy Hemsworth Architecture

9781733754651

AUTHORS

Dave Atkins, Treesource

Roy Anderson, *Vice President*, The Beck Group

Emily Dawson, *Partner, Principal Architect*, Kaiser + Path

Lech Muszynski, *Professor,* Department of Wood Science and Engineering, Oregon State University

CONTRIBUTORS

Bryan Beck, *President*, The Beck Group

Ben Kaiser, *Owner*, Kaiser + Path

Craig Rawlings, *President & CEO*, Forest Business Network

The work upon which this document is based was funded in part through a cooperative agreement with the US Forest Service, Wood Innovations Grants. The US Forest Service is an equal opportunity provider, employer, and lender.

DISCLAIMERS

Where applicable in this report, column and row data may not sum to totals and subtotals due to rounding. This variation may also appear in the corresponding analysis where it references data in a table, chart, or other graphic.

Every effort was made to present accurate information from the best sources available, but the authors make no warranty about the completeness, reliability, or accuracy of this information. Any action taken based on information in this report is strictly at your own risk—the authors and contributors will not be liable for any losses or damages in connection with the use of this report.

ON THE COVER

UPPER SKEENA RECREATION CENTRE, HAZELTON, BC

Source: Hemsworth Architecture; Photo Credit: Ema Peter Photography

WHEN ITS 44-YEAR-OLD ICE ARENA was condemned because of safety concerns, of the community banded together to support construction the Upper Skeena Recreation Centre. It's built of heavy timber and wood-frame construction, and it delivered efficiency, innovation, and hope to this remote northern region.

- To speed construction, wood wall and roof panels were prefabricated and installed on-site.

- Local residents were employed to build the panels and help erect the frames; the walls and the roof were built with lumber and plywood manufactured by builders in the community.

- The practical yet striking wood structure was made possible through grassroots community involvement and gives the region a long-lasting focal point of pride.

LOCATION: HAZELTON, BRITISH COLUMBIA

SIZE: 5,050 SQUARE METERS

COMPLETION: 2019

ARCHITECT: HEMSWORTH ARCHITECTURE

STRUCTURAL ENGINEER: EQUILIBRIUM CONSULTING

CONSTRUCTION MANAGER: YELLOWRIDGE CONSTRUCTION LTD.

WOOD SUPPLIER: STRUCTURLAM MASS TIMBER CORPORATION

OWNER: REGIONAL DISTRICT OF KITIMAT-STIKINE

SPECIES: WESTERN RED CEDAR, DOUGLAS FIR, BIRCH

PROJECT MATERIALS: GLULAM, LUMBER, PLYWOOD, SOLID-SAWN HEAVY TIMBER

STRUCTURAL SYSTEMS: LIGHT FRAME, LOW-RISE, PANELIZED, POST AND BEAM, PREFABRICATED

COVER PHOTO CREDIT: EMA PETER PHOTOGRAPHY, COURTESY HEMSWORTH ARCHITECTURE

SPONSOR SPOTLIGHT: *A special thanks goes to 2022 report sponsors Weyerhaeuser, DCI Engineers, and WholeTrees, without whose support the report would not be possible.*

TABLE OF CONTENTS

CHAPTER 1: INTRODUCTION _____ 1

1.1   Why a Mass Timber Report? _____ 1
1.2   What Is Mass Timber? _____ 2
1.3   How Is Mass Timber Used? _____ 9
1.4   Defining the Mass Timber Supply Chain _____ 10
1.5   Measurements and Conversion Factors _____ 10

CHAPTER 2: THE FOREST RESOURCE _____ 16

2.1   Characterizing the North American Forest Resource _____ 16
2.2   Forest Sustainability _____ 23
2.3   Forest Diversity _____ 32
2.4   Forest Health _____ 32
2.5   Forest Fire Resilience _____ 33
2.6   Forest Carbon _____ 35
     CASE STUDY: FRERES LUMBER _____ 40

CHAPTER 3: RAW MATERIALS _____ 42

3.1   Raw Material Specifications _____ 42
3.2   North American Lumber Supply _____ 51
3.3   The Mass Timber Industry's Estimated Demand for Raw Materials in 2021 _____ 61
3.4   Supplying the Mass Timber Market: Sawmiller Perspectives _____ 62
3.5   Carbon Considerations _____ 64

CHAPTER 4: MASS TIMBER PANEL MANUFACTURING _____ 67

4.1   Mass Timber Panel Types _____ 67
4.2   Mass Timber Panel Manufacturing Process Descriptions _____ 68
4.3   North American Mass Timber Plants _____ 72
     CASE STUDY: TIMBER AGE SYSTEMS _____ 74
4.4   Mass Timber Manufacturers: Company and Facility Details _____ 79
4.5   North American Mass Timber Manufacturer Services _____ 79
     CASE STUDY: EUCLID TIMBER FRAMES _____ 84

CHAPTER 5: DESIGNERS & SPECIFIERS _____ 88

5.1   Carbon Considerations _____ 88
     CASE STUDY: PORTLAND INTERNATIONAL AIRPORT MAIN TERMINAL EXPANSION _____ 96
5.2   Elements of Design _____ 100
     CASE STUDY: PMX 15 _____ 110
     CASE STUDY: HESS TIMBER/GOOGLE _____ 127
5.3   Project Management and Coordination _____ 129
5.4   Building Codes _____ 132
     CASE STUDY: BRITISH COLUMBIA _____ 137

CHAPTER 6: BUILDERS _____ 139

6.1   Market Context _____ 140
6.2   Wood as a Construction Material _____ 140
6.3   The Mass Timber Building Experience _____ 143
     CASE STUDY: IDAHO CENTRAL CREDIT UNION ARENA _____ 156
     CASE STUDY: VAPROSHIELD _____ 162
     CASE STUDY: SOPHIE RADICH SCHOOL _____ 166
6.4   Quantifying Cost Savings _____ 169
     CASE STUDY: MONTANA FWP CONFERENCE CENTER ADDITION _____ 170

CHAPTER 7: OCCUPANTS _____ 173

7.1   Health _____ 173
7.2   Comfort _____ 176
7.3   Behavior _____ 180

CHAPTER 8: OWNERS AND DEVELOPERS _____ 183

8.1   Carbon Considerations _____ 183
8.2   Market Development: US Mass Timber Projects _____ 187
8.3   Rationale and Motivation _____ 190
     CASE STUDY: PAE LIVING BUILDING _____ 192
8.4   Executing an Innovative Project _____ 201
     CASE STUDY: VAAGEN _____ 208

SPONSOR **SPOTLIGHT**

DCI Engineers has been providing client-focused wood and mass timber services for over three decades. With 350+ employees in 14 offices, DCI is licensed in all U.S. states and most Canadian provinces. Milestone mass timber projects have included: The Bullitt Center in Seattle, which is the first Living Building and greenest office project in the U.S.; 1 De Haro, an office/light industrial building that is the first Cross Laminated Timber (CLT) of its type in California and the first CLT building in the U.S. delivered by rail; Project One, the first in the country to utilize Mass Plywood Panels (MPP); and several First United Bank, Net Zero buildings across Texas, the first projects in the U.S. to utilize Southern Yellow Pine CLT. DCI, a member of the SE 2050 Commitment, recognizes mass timber as a low carbon alternative and is committed to green building design.

To learn more about what sustainable measures DCI is taking on their projects, visit: _dci-engineers.com_.

THE BULLITT CENTER | SEATTLE, WA

1 DE HARO | SAN FRANCISCO, CA

PROJECT ONE | OAKLAND, CA

FIRST UNITED BANK | TX & OK

Watch the story of 1 De Haro, San Francisco's first Cross Laminated Timber building and the benefits of mass timber construction. From sustainability to constructability, mass timber is breathing new life into one of the industry's oldest materials—wood.

WATCH NOW! 1 DE HARO: A CASE FOR MASS TIMBER
https://vimeo.com/480069029

PUBLISHER'S MESSAGE

The 2022 version of the *International Mass Timber Report* is the fourth iteration of our comprehensive guide to all things mass timber. Each year we look for ways to make it better. In 2021 we added "International" to the title and focused on bringing international content to the report, reflecting the true global nature of the mass timber movement.

Return readers will notice even more changes in 2022 as we continue our focus on continually improving the report and ensuring the content remains relevant year after year. The core information remains in Chapters 1–8, covering everything from "What is mass timber?" through sourcing the wood and production of the various types, to issues related to design, engineering, and building codes, and wrapping up with a focus on builders, occupants, and owners and developers.

Added features in 2022 include in-depth articles on what we think are some of the hottest topics facing the industry. In this report you will find our first feature story on carbon, titled *All Models Are Wrong; Some Are Useful.* This article by longtime industry expert Dave Atkins takes an unvarnished and unbiased look at carbon accounting and Life Cycle Analysis (LCA) tools, important factors in understanding the value of mass timber in fighting climate change.

Our second feature story is on modular and off-site construction: *Mass Timber Systems: From Building Products to Product Platforms.* Modular components are becoming a more important piece of the puzzle as the industry seeks ways to make mass timber even more cost-effective.

We have also developed *The Mass Timber Interview,* a forum where we share the thoughts of industry leaders. Finally, we have created an industry tracking tool, *The Mass Timber Performance Index,* to share mass timber capacity, production, and cost information from around the world—vital information that can help you understand where the industry is today and where it is going.

We hope you will also find the report a little easier to use with the addition of case study and advertiser indexes in the back of the book.

Speaking of advertisers and sponsors, we want you to know that this report wouldn't be possible without their support. We thank them for that, and we encourage you, our readers who might need mass timber-related products or services, to check them out.

Moving forward, we invite your input on ways we can continue to improve the report. We want to make sure we are delivering the information you need and find valuable. What are we missing? What can we do better? We are always interested in new developments, designs, projects, and ideas. You can reach us easily through *www.masstimberreport. com.* In addition, there's a call for submissions on the authors page that will reach me directly.

Best,

Dave

David Parcell
Publisher

THE ORIGINAL MASS TIMBER:

DO MORE FOR THE PLANET BUILDING WITH STRUCTURAL WOOD

The appeal of mass timber lies not just in its aesthetics, biophilic properties and structural integrity, but also in the role it can play in fighting climate change.

WholeTrees Structures and Port Blakely have formed a unique partnership to harness the power of carbon forestry in the built environment. They are bringing a new generation of the original mass timber – Douglas fir Structural Round Timber (SRT) and Sawn Heavy Timber (SHT) – to the market.

Sourced from Port Blakely's 10,000-acre Winston Creek Carbon Forest, Douglas fir SRT and SHT are competitive with engineered wood on both price and structural capability. Fabricated to order, they are delivered with full transparency from forest-to-building, offering specifiers a means by which to expand upon the carbon benefits of their mass timber buildings.

The Winston Creek Carbon Forest is the largest privately-owned forest in the Pacific Northwest to be certified by the American Carbon Registry. Port Blakely has made a commitment to manage Winston Creek on a 60-year harvest rotation, up to 25 years longer than the industry norm. This will result in increased net carbon capture along with ecosystem benefits including improved biodiversity, wildlife habitat, and water quality.

Thanks to exceptional forestry practices and a commitment to a 60-year harvest rotation, Port Blakely's Winston Creek Carbon Forest is the largest privately held forest in the Pacific Northwest to be certified by the American Carbon Registry for additive carbon sequestration. Structural wood products sourced from Winston Creek and other Port Blakely forests, available through WholeTrees Structures, allow architects and other specifiers to expand upon the climate benefits of their wood buildings.

This new generation of the original mass timber offers the ideal climate solution for the built environment.

Learn how you can do more for the planet at wholetrees.com/port-blakely or info@wholetrees.com.

Port Blakely is a family-owned sustainable forest products company with forestland in Washington, Oregon, and New Zealand. Whole Trees is a rapidly growing innovative building products service provider and fabricator. Together, they are bringing Carbon Forest Sourced Douglas fir Structural Round Timber (SRT) and Sawn Heavy Timber (SHT) – the Original Mass Timber – to the commercial construction market.

ALL MODELS ARE WRONG, BUT SOME ARE USEFUL

DAVE ATKINS

An elephant walks into a bar, where an architect, a forest owner, a developer, a tech sustainability manager, and the head of an environmental organization are having a drink. They are all blind. Each encounters a different part of the elephant: one the trunk, one the tail, one the ear, one the leg, and one hears it trumpet. When asked to describe the elephant, the one who encounters the trunk says it is very similar to a boa constrictor, the tail generates the description of a thick rope, the ear is described as a large fan, the leg is likened to a tree, and the listener says it must be a tugboat.

Sustainability is the elephant in the room that we all want to achieve. We all, however, have our own ideas about what that looks like. Each of the people in the bar gave an accurate description of the part of the elephant they encountered, but without working together and sharing information, none of them completely understands what it is.

We all know about the dire warnings issued by the Intergovernmental Panel on Climate Change (IPCC) and the speeches given in Glasgow at the United Nations Climate Change Conference (COP26). The business world has weighed in: investment and insurance firms have expressed grave concerns about the effects of climate dis-

ruption on economic values, and the US Chamber of Commerce stated about a year ago that climate change is happening, and it is caused primarily by humans. The chamber also said that businesses are essential to solving this problem and that a market-based policy solution is needed to drive businesses to change their behavior.

So, how do we transition to a low-carbon society?

In the built environment, a variety of certification systems and modeling tools have been developed to support creation of more sustainable buildings, providing tremendous examples of net-zero buildings. These certification systems include Leadership in Energy and Environmental Design (LEED), Green Building Initiative (GBI), and the Living Building Challenge (LBC). These systems have shown us what is possible and have created

standards and guidelines to steer users through the process. Models have been developed to assess the carbon footprint for the use and maintenance of buildings, and, in more recent years, much of the focus has been on embodied carbon content. That has been accompanied by the recognition that a rapid reduction in emissions over the next 8 years to achieve the goal of a ~50 percent reduction by 2030 and net zero by 2050, must be led by reducing embodied carbon. They are sunk costs once expended, whereas the use and maintenance carbon can be ameliorated by the continuing shift away from fossil energy sources.

The desire to change the embodied carbon has driven the excitement about and expansion of the use of wood in place of fossil-energy-intensive materials, given the carbon storage wood provides and the reduction of greenhouse gas (GHG) emis-

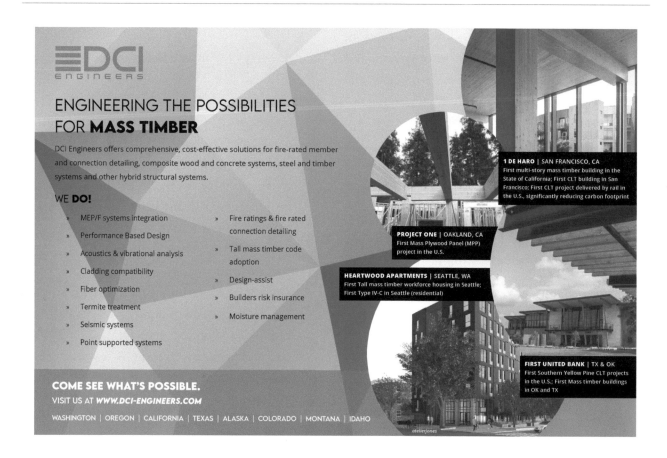

sions from comparative manufacturing processes for materials such as aluminum, steel, and concrete. Life Cycle Inventories and Analyses (LCI/LCA) and modeling tools that allow architects, engineers, and others to make comparisons have proliferated. Along with those have come debates about the appropriateness of the assumptions that have gone into some of the modeling efforts at the building level, but more so in the sourcing of wood and the management of forests.

We explore some of these critiques (with links/references) here, and reflect on how the critics' view of the elephant can slant their perspective of what "sustainable" means. We ultimately anchor back to our title: "All Models Are Wrong, But Some Are Useful." This essay will focus on both parts: "some are useful" and "all models are wrong," as we work to avoid having the perfect become the enemy of the good. This essay does not pretend to be the final word on carbon in wood and forests and their role in helping solve climate disruption; rather the goal is to provide information and context that will help us all better describe the elephant and continually improve.

MODELING TOOLS FOR THE BUILT ENVIRONMENT: LCI/LCA

Life Cycle Inventories (LCI) contain the data that underlies the Life Cycle Analyses (LCA). LCAs are used to examine the environmental effects of different materials, including their Global Warming Potential (GWP) and their effects on water, eutrophication, ozone, and more. They are the basis for the Environmental Product Declarations (EPD) that are used to compare products. There are international standards (ISO *14044, 21930*) that govern the processes to be followed to ensure standards are consistent and allow for compari-

sons, just as the various building and forest certification systems have criteria and standards that provide accountability through verification and modeling for transparency and credibility. These sources provide transparent information for people working in the forest and in the built-environment sectors so they can understand the sources of their material and some of the environmental consequences of its use.

A suite of tools has been developed in recent years to help designers, developers, engineers, and contractors make choices among products, assemblages, and Whole Building Life-Cycle Analysis (WBLCA). These tools are useful at various stages, from conceptual design to detailed planning to procuring the materials for the built environment. The tools include open source nongovernmental organizations (NGOs), university- and government-supported tools, and proprietary company-built tools.

Good sources for these models include these:

- The *Carbon Leadership Forum* (CLF): A collection of architects, engineers, contractors, material suppliers, building owners, and policymakers who are taking bold steps to decarbonize the built environment, with a keen focus on eliminating embodied carbon from buildings and infrastructure.

- *Consortium for Research on Renewable Industrial Materials* (CORRIM): It is made up of a multitude of universities, government groups, and trade associations. CORRIM conducts and manages LCA research on the environmental impact of production, use, and disposal of forest products.

- *Athena Sustainable Materials Institute*: A Canadian membership-based nonprofit research collaborative bringing LCAs to the construction sector.

Other proprietary tools have been developed by individual firms.

This list has been rapidly growing and changing as new information and products are developed and evaluated, and these sites help the industry stay current. The LCI, the database of information used to conduct the assessments, is vital to the process.

As with any human endeavor, people have different perspectives on exactly how the work should be done and support their favorite approach.

Sustainability can be especially challenging because it requires finding the sweet spot where solutions work socially, environmentally, and economically, and doing so requires tradeoffs. If you are a publicly traded timber company, you are holding a different part of the elephant than if you are an environmental NGO, or if your focus is on labor, local residents, and/or indigenous rights.

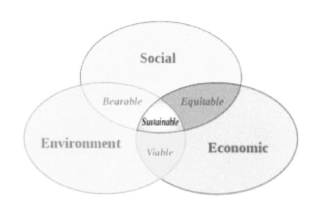

FIGURE 1: SUSTAINABILITY VENN DIAGRAM

operating in this arena as their staffs can move more quickly through the decision-making process without running the model every time. Studies published in January 2022 as part of a special issue of *Sustainability* offer examples from China, Chile, Austria, and the US. The results varied as a result of their location but all showed a substantial reduction in embodied carbon.

Comparing materials and methods: Whether comparing different structural components or insulation choices for specific applications, WBLCA can facilitate the evaluation of alternative designs. **Figure 4** compares the carbon footprint of wood fiber insulation with fiberglass, mineral wool, spray foam, and extruded polystyrene.

HOW ARE THESE TOOLS USEFUL?

Education and awareness: Whether targeting practitioners, clients, investors, government officials, or the general public, models are great tools to show people why it is important to consider embodied carbon and how it is accounted for. A valuable result of practitioners' repeated use of the models is the development of an intuitive understanding of the choices involved and their relative magnitude. The development of this intuitive understanding is extremely valuable to firms

Accountability: Finally, the models and their outputs provide methods of verification and accountability, demonstrating what has been accomplished or what could be accomplished with new policies. This is vital to meeting the requirements of various voluntary certification systems. But it is also becoming important to meet regulatory requirements, as some local governments are establishing embodied energy standards for materials and buildings. The voluntary carbon offset

Phase 1: Comparative whole-building LCA

Phase 5: Integration & communication

Phase 2: Regional demand assessments (China, USA, Europe, Southern Cone of South America)

Phase 4: Regional forest modelling

Phase 3: Global trade modelling

FIGURE 2: FIVE PHASES OF GLOBAL MASS TIMBER IMPACT ASSESSMENT
Source: https://www.mdpi.com/2071-1050/14/2/758/htm

sector is expected to grow as a result of agreements made in Article 6 at COP26, based on the Task Force on Scaling Voluntary Carbon Markets (TFSVCM). Monetary rewards could result from reducing the embodied carbon in buildings. As a quantifiable, verifiable carbon savings technique accomplished at the time of construction, this could help accelerate the movement to low-carbon material choices. A crucial element of this effect will be the price of the avoided carbon, which has been increasing.

The Nature Conservancy initiated, with a variety of partners, a huge research effort to examine the potential for mass timber use in buildings world-wide to reduce carbon emissions and store carbon in the built environment. It is also examining the potential for a sustainable supply of wood to meet the potential increased demand.

Some results of the first three phases were published in the special issue of *Sustainability* in January 2022. LCAs were conducted to compare buildings in three parts of the world: China, the US, and Europe. The bottom line was the use of mass timber was beneficial in all locations. Analysis of the US's ability to meet demand from existing forests on a sustained basis was validated, based on the US Forest Service's Forest Inventory of plots across all ownerships. The remaining research to be done by 2023 will complete the regional forest modeling, and, most importantly integrate the results of all phases. The Nature Conservancy is working on what it calls the 3S Framework (see sidebar) modeling, the carbon effects from the forest to the building. The three S's are sequestration of carbon in the forest, storage in buildings, and substitution of lower embodied carbon materials for higher.

HOW ARE THESE TOOLS WRONG?

All models are built upon assumptions and data sets, as well as the current state of product availability. The data sources can never be complete, given the variation in companies, their manufacturing processes, differences in transportation, and sources of raw material. Assumptions about the longevity of the building, the method of its disposal, and the potential for reusing materials versus landfilling can all influence the results. However, good data collection, management, and use can minimize the errors inherent in the modeling process. The largest point of contention surrounds the management of the forests and whether sustainable harvesting and conversion of trees to wood products that can substitute for carbon-intensive materials is better than letting the forests grow and continue to store more carbon. The assumptions made when approaching this fundamental question are crucial.

The Forests Dialogue (TFD), an effort to address a variety of international forest issues, is hosted by the Yale School of the Environment. They took on the issue of Climate Positive Forest Products (CPFP), focusing on mass timber and the potential growth of this sector worldwide, and the potential benefits and pitfalls. A valuable background paper prepared for a TFD event in April 2021 provided an excellent synthesis of the areas of agreement and disagreement around the expanded use of mass timber in North America, Europe, and elsewhere in the world. We have drawn upon that, along with other sources of information, to provide you with some insight into this topic. For an in-depth exploration, access the background paper and the *Co-Chairs Summary Report* that also identifies next steps. Keep in mind that TFD is a global dialogue, and the issues it raises vary depending on the region under discussion. The initial meeting showed the value and importance of keeping the dialogue moving forward as the discussions revealed many folks had a hold on different parts of the elephant, and the dialogue helped all the participants to better understand what the whole elephant looks like.

If the models are looking at long time spans, the validity of their assumptions become more tenuous than in the near term. What follows is an exploration of some of the critiques of LCAs related to mass timber, the places they can potentially go wrong, and the ramifications of those errors.

One critique relates to the substitution benefits of using wood versus fossil energy in calculating the carbon content of building materials. Graphics such as **Figure 3** show that the substitution benefits are unchanged over long periods. Critics point out that this doesn't reflect likely changes in the carbon content of the alternative materials. In fact, the whole purpose of COP26 and the LCAs is to ensure our relationship to fossil carbon changes.

Nonfossil energy production is increasing worldwide, changing the carbon content of alternative materials over the next few years and decades. Ongoing research and development work to re-

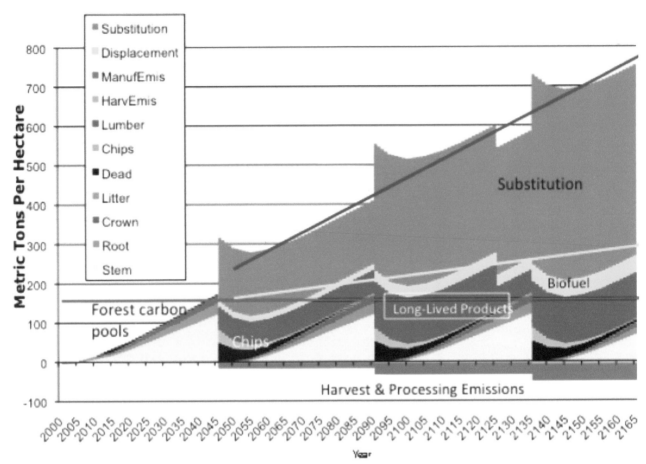

FIGURE 3: SUBSTITUTION BENEFITS
Source: CORRIM Factsheet #5

duce the carbon dioxide intensity of concrete is in the works and has been done. The use of nano-cellulosic wood in concrete is reducing emissions by 15 percent to 50 percent. The first use of it in a bridge in California has been demonstrated. Using biochar as an additive significantly reduced emissions in research trials. Adoption and acceptance of new technologies may take at least a decade, given that it took more than a decade with mass timber. So the critique of the graph is valid in the timescale. The substitution benefits in the buildings completed in the past 10 years, however, are very real and can be expected to be similar in the next 5 to 15 years. This is substantiated by the many LCAs completed and published in many parts of the world. This time frame is of utmost importance for meeting our GHG emission goals worldwide.

The critics of the substitution benefits generally don't point out that in addition to changes in the carbon content of fossil-energy-intensive materials, the use of wood residues is changing. For long-lived projects, 56 percent of the log is residue rather than lumber. Some of the wood residue goes into fiberboard products that can have long lives in furniture and other composite materials. But with the development of wood fiber insulation, as at the GO Lab plant in Maine that is under construction, we can see residue not only

CARBON FOOTPRINT

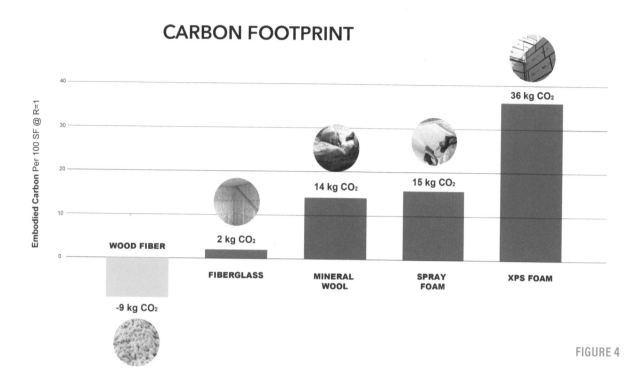

FIGURE 4

from manufacturing but also the potential for residue from harvesting becoming feedstocks for long-lived products in the built environment.

Other new products created from residue are products from Made of Air. They are using wood residue-based biochar (90 percent) and sugar cane

residue (10 percent) to make bio-based thermoplastics. These are used in a variety of products ranging from sunglass frames to building cladding. The biochar carbon is captured for centuries to millennia. The process of making the biochar produces syngas that can be used for bioenergy. The result is called BioEnergy Carbon Capture and Storage (BECCS). Incorporating the char into structures made from mass timber further adds to the long-term storage of carbon derived from natural carbon solutions and is another step toward making cities carbon sinks.

In carbon discussions, people often get fixated on the long-lived product, discounting the carbon value and importance of short- to medium-term products. Even when these products are relatively short-lived, they are displacing materials made from petrochemical plastics that have a high embodied carbon content. The production and use of paper, cardboard, and other packaging from a renewable wood residue source that can be recycled multiple times, is biodegradable, and can be returned to the soil are all important to a low fossil-carbon circular economy. Cardboard is recycled at about a 90 percent rate, and the fibers can be reused 5 to 7 times.

The critique showing new substitution benefits 50, 100, and 150 years out is valid, but it doesn't mean anything in terms of the actual benefits achieved next year or likely 10 years from now. The take-home point is that substitution benefits are real in the near term, and that is what counts.

Another critique relates to the carbon differential between harvesting and using the wood versus continuing to let the forest grow, but the "let-it-grow" advocates typically ignore natural disturbances when they show the carbon storage line steadily increasing. Wildfires, windstorms, and insect and disease outbreaks occur and dramatically change the ability of the forest to capture carbon dioxide and release substantial amounts of carbon. A combination of disturbance events, such as 2 fires within 10 to 30 years of each other or an insect epidemic followed by a wildfire, can cause significant fluxes of carbon to the atmosphere. This was evidenced by repeated fires in large wildernesses and Yellowstone National Park, for these areas have been managed to allow natural processes to operate as much as possible. It is essential, therefore, to make these discussions specific to forest ecosystem and region.

In forests with much longer intervals between natural disturbances, the risk of carbon loss is lower and the potential for growing more wood and capturing more carbon for use in buildings is very real. A research paper from 1995 showed that coastal Douglas fir, when commercially thinned, can extend high growth rates out to

117 years if no disturbances occur. Thus, there are ways to achieve more growth and to continue to harvest logs. It doesn't have to be an either/or proposition. We can have more wood *and* more carbon stored in the forest. We are already seeing the natural disturbance regimes being altered by climate change, however, as evidenced by the 2020 wildfires in western Oregon.

The let-it-grow approach is also dependent on discounting the product substitution benefits, which, as discussed earlier, is not reasonable for the next decade or two.

Another critique addresses the starting point of the forest and carbon stocks at the time of harvest. Are you starting with an old-growth forest in the Pacific Northwest at the time of harvest or a forest that is on a sustained harvest on a cycle of 40 to 80 years? Old-growth forests have substantially higher carbon stocks at the initial harvest and thus will have an extended period where carbon is released from the large accumulation in the downed logs, litter, and duff. The result is a continual release of carbon greater than what the new forest will capture in its next rotation. With established, sustained rotations, the additional carbon release is not an issue. Knowing the source of the wood can be important, but in the US, the harvesting of old growth is minuscule. The vast majority of softwood lumber comes from private lands in which old-growth forests, if they were such, were harvested 1 to several cycles ago.

This discussion gets conflated when old growth is used as a synonym for primary (or virgin) forest—a forest that hasn't been harvested before. This assumption is problematic in many forests where the natural disturbance regime can mean very little carbon is stored beyond the existing

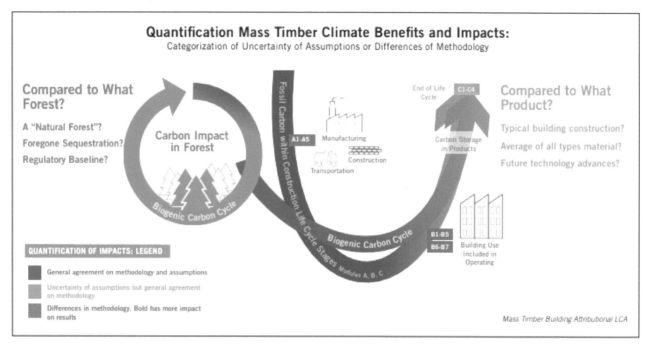

FIGURE 5: QUANTIFICATION MASS TIMBER CLIMATE BENEFITS AND IMPACTS

Source: TFD background paper

trees. Again, the disturbance regimes are an important factor to consider.

Climate Smart Forestry (CSF) or Improved Forest Management (IFM) are other terms that are surfacing more often and can mean different things to different people. These terms are often used when referring to management techniques that can capture more carbon and make more wood available. These techniques can include controlling competing vegetation, including other trees, to enhance the growth rate of desired trees that can capture more carbon and can effectively be turned into long-lived wood products. The American Forest Foundation and The Nature Conservancy have initiated the Family Forest Carbon Program, which is helping small landowners access carbon-offset markets by applying one of these practices so they can get paid to store more carbon and coincidentally have more wood to harvest at the end of their contract. At COP26,

Article 6 was adopted to incorporate rules for the application of the transparent methods developed by the *Task Force on Scaling Voluntary Carbon Markets (TFSVCM)*. Natural carbon storage is part of this article. The opportunity to leverage corporations, governments, and more to invest in improved forest management for carbon benefits is significant.

Many people also use these terms when describing "defensive carbon management." This involves helping the forests adapt to shifting climatic conditions, including more drought, more fire, more insect attacks, and so on. Many Western forests—especially public ones but private as well—are too crowded, making them vulnerable to wildfire, defoliating insects, and bark beetles. The carbon (trees) needs to be removed, leaving the species that are more resistant to drought, fire, and insects. Making these forests more resilient can help them continue to capture and store carbon and provide more wood in the future, but the immediate effect is to draw down the stocks of carbon.

What is the point of illustrating all these potential and real errors associated with LCA, building models, forest models, and the like? The graphic above (**Figure 5**) shows us the greatest point of difference is over forest management, and it needs the most attention to effectively move forward. It is important for us all to step back from the details of forest modeling and try to see the whole elephant. If we myopically focus only on carbon, we are likely to cause other problems, whether they be with our watershed values; our biodiversity values; our economic values; or our esthetic, spiritual, and recreation values. Forests provide multiple benefits; some are managed for a sustainable supply of wood, and others are managed for special features, including biodiversity or recreation.

Many forests are managed for a blend of these values; they provide some wood, but that is not their primary purpose. They might be a crucial watershed for an urban population or for agricultural irrigation, or they might be primarily recreational. Forests are not a binary choice. In the US, 38 percent of the forestland is owned by "family forest owners," a group with as diverse a set of management goals and objectives as you can imagine. Bringing them into the mix for management of carbon, wood, wildlife, water, recreation, and resilience to climate change is vital. Only ~18 percent of forest land is corporately owned and dedicated primarily to wood production.

We can focus on what is wrong with models and argue specific points of disagreement, or we can realize forests and all their values are important

and use the models to guide our discussions, identify the sticking points, and work to understand one another's perspectives and reasons for skepticism. We can then work to bring new and better information to the dialogue so that we can build trust.

Yes, all the models are wrong, but they are useful in helping us confirm the importance of efficient, efficacious management throughout the whole supply chain, from the diversity of forest purposes to constructing the building to deconstructing the building for material reuse. The continual striving for improvement of the data, the models, and the management are essential to achieve our goals. The *Climate Smart Forest Economy Program* is an effort to help us all work in this direction.

A final closing thought is that, if we don't get lost in the details, the models can help us think through and discuss all the different aspects of the sustainable elephant. The models can illustrate different scenarios, pathways, and alternate choices to better help us understand the ramifications. It is essential that the collection of interests, with all our different areas of expertise, share those perspectives. That way, we can reach the shared goal that each of us holding a part of the elephant will listen and envision what is being shared and gain a better understanding of the shape and contours of this giant, called sustainability, that is key to our collective success.

3S FRAMEWORK FOR FORESTS: *An Overview*

Forest protection, restoration, and management are increasingly being recognized as important actions to mitigate climate change. But crucial gaps still remain in our understanding of the full opportunities forests can provide: as carbon sinks that require conservation and restoration; and as sources of sustainable forest products (e.g., mass timber as a building construction material) that store carbon and substitute for fossil-based materials. A holistic perspective is needed to support policymaking, investment decisions, and actions that balance the function of forests and sustainable forest products for climate, environmental, and social benefits.

Forest systems and sustainable forest products—specifically wood for use in construction—are complex systems. Linking these two systems involves a creating a system in itself that does not now exist. Forestry practices are not accounted for, for example, when assessing the climate mitigation potential of the wood products used in construction. Carbon tends to be expressed only through building materials, and the impacts are widely misrepresented until building materials get to site.

We are, therefore, developing the 3S Framework for forests: a tool for stakeholders across multiple sectors to assess how their choices can maximize the climate change impact of forests and sustainable forest products. The tool will help compare different scenarios in terms of carbon absorption and sequestration (the sink function); carbon storage (the biocarbon stored in forest-based products); and carbon substitution (the fossil carbon emissions avoided).

The 3S Framework is a robust analytical tool that can calculate carbon impacts and allow various stakeholders to assess how their choices can maximize the potential of forests and forest products to mitigate climate change. The 3S Framework provides a snapshot of carbon cycling between buildings and a forest ecosystem. These are also the major components that make up the 3S Framework. In its current version, the framework quantifies the storage of carbon in cities (e.g., buildings) or ecosystems (e.g., trees), in addition to the transfer of carbon from a forest ecosystem to a building. It also simulates associated emissions of carbon into the atmosphere from manufacturing and construction of timber buildings and their steel/concrete counterparts, enabling comparison of respective carbon emissions.

We are developing this framework and are eager for input from across the stakeholder spectrum. We'd like to know:

- *Does this tool sound like something that might be useful for you?*

- *If no, why?*

- *If yes, how would you use the tool, and how does it fill gaps in your knowledge of your project? Please proceed to the following to complete a short survey:*

We are in the final stages of developing and testing the tool on a range of case studies. The prototype of the framework will be completed in autumn of 2022. If this tool sounds of interest, and you'd like to discuss your feedback further, please get in touch with Rachel Pasternack, *rachel.pasternack@tnc.org*.

The 3S Framework is a collaborative project developed by The Nature Conservancy and Dr. Galina Churkina, in partnership with the Climate Smart Forest Economy Program (CSFEP). This project is supported by the Good Energies Foundation. ⬤

MASS TIMBER PRICE INDEX

ROY ANDERSON,
Vice President, The Beck Group

Published prices are available for a variety of building products, including steel, concrete, lumber, plywood, oriented-strand-board (OSB), and more. In contrast to these well-established price reporting services, no such service is available for mass timber in North America. The following information, therefore, illustrates the estimated average value of 3-ply Cross-Laminated Timber (CLT) panels Free on Board (FOB) the manufacturer's plant between 2016 and 2021. As the data shows, for a period of about four years, the estimated price of CLT ranged from about $20 per cubic foot to $25 per cubic foot. Beginning in the second quarter of 2020, however, the price skyrocketed, reaching more than $45 per cubic foot in early 2021. The prices are reported on a quarterly basis between January 2016 and December 2021. Note that the values on the left axis are the prices expressed in $ per cubic meter, and the values on the right axis are the prices expressed in $ per cubic foot.

MASS TIMBER PRICE INDEX DISCUSSION

As **Figure 1** illustrates, the estimated price of CLT panels slowly ramped up through mid-2018 and then dropped slightly until Q2 2020. Still, over that period prices were relatively stable, ranging from about $20 per cubic foot to nearly $26 per cubic foot. Then in mid-2020, the COVID-19 pandemic increased demand for wood products as many homeowners adapted their living spaces to better accommodate working from home and took on improvement projects to make spending more time at home more enjoyable. At the same time, wood products manufacturers were unable to increase production enough to meet increased demand—primarily because of staffing shortages. As a result, softwood lumber prices spiked to unprecedented levels by mid-2021. Because softwood lumber is a key raw material in most types of mass timber panels, the prices of mass timber products rose in a similar fashion.

The values in **Figure 1** are not panel sales prices reported through surveys or transaction evidence. Rather, they are calculated from a financial model of a mass timber manufacturing plant. The model includes the costs of production (e.g., labor, adhesives, supplies, repairs and maintenance, utilities, etc.), raw material (e.g., lumber), and a profit and risk allowance for the mass timber manufacturer. Because the prices are estimated from a model that uses a cost-plus profit and risk approach, there may be times when other market supply and demand factors cause actual market prices to deviate from those modeled. The information nevertheless provides a reasonable estimate of basic mass timber value.

NORTH AMERICAN MASS TIMBER: LUMBER USAGE, MASS TIMBER CONSUMPTION, AND MASS TIMBER IMPORTS

North American mass timber manufacturing plants' rate of production is another topic of great interest to those who follow the industry. Like mass timber market prices, little solid information is available directly from manufacturers about their production capacity and production

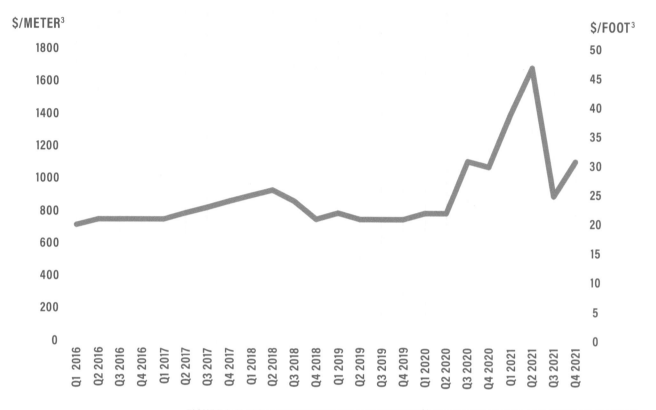

$/METER³

1800
1600
1400
1200
1000
800
600
400
200
0

$/FOOT³

50
45
40
35
30
25
20
15
10
5
0

Q1 2016 Q2 2016 Q3 2016 Q4 2016 Q1 2017 Q2 2017 Q3 2017 Q4 2017 Q1 2018 Q2 2018 Q3 2018 Q4 2018 Q1 2019 Q2 2019 Q3 2019 Q4 2019 Q1 2020 Q2 2020 Q3 2020 Q4 2020 Q1 2021 Q2 2021 Q3 2021 Q4 2021

FIGURE 1: MASS TIMBER PRODUCT PRICE INDEX ($/METER³ LEFT AXIS AND $/FT³ RIGHT AXIS)

rates. This section nevertheless estimates North American mass timber production and lumber use from 2019 to 2021.

As shown in **Table 1**, an average of roughly 300,000 cubic meters of mass timber products were used in building construction each year. The table also shows that during that time, mass timber building construction consumed roughly a quarter billion board feet of softwood lumber per year. During that same period, an average of about 20,000 cubic meters of mass timber products were imported into North America each year. Thus, after accounting for the volume of mass timber imported, North America's mass timber plants that are focused on making panels for use in building construction have been operating at

rates estimated to be between 50 percent and 67 percent of their practical capacity.

NORTH AMERICAN MASS TIMBER LUMBER USAGE AND CAPACITY DISCUSSION

As the data in the **Table 1** shows, the number of publicly reported mass timber buildings in the US and Canada has grown year over year between 2019 and 2021. The number of new buildings, however, has not grown at the rates once projected for the mass timber marketplace—likely due largely to COVID-related market uncertainties and constraints. See Chapter 8 for additional market analysis. One key point made there is that the number of mass timber projects in the planning

YEAR	Publicly Reported Number of Mass Timber Buildings Constructed in US & Canada	Estimated North American Use of Mass Timber Products (Cubic Meters per Year)	Estimated Board Feet of Lumber Used in mass timber in North America (MBF Lumber per Year)	Estimated Imports of Mass Timber Products into the US (Cubic Meters per Year)	Apparent North American Mass Timber Production (Cubic Meters per Year)	Apparent North American Mass Timber Production (Cubic Meters per Year)	Estimated Percent of Practical Building Panel Mass Timber Manufacturing Capacity Utilized
2019	143	282,900	224,500	15,000	267,900	400,000	67%
2020	168	303,500	241,300	24,700	278,800	541,000	52%
2021	170	323,800	257,300	19,300	304,500	520,000	59%

TABLE 1: ESTIMATED NORTH AMERICAN MASS TIMBER LUMBER USAGE AND PRODUCTION (2019 TO 2020)

stages continues to grow. In late 2020, a total of about 600 projects were in the design phase; as of late 2021, that number had grown to 700.

Also, in reference to the analysis presented here, the Softwood Lumber Board commissioned a *Mass Timber Outlook*[1] study that was published in October 2020. A key finding was that by 2025, softwood lumber usage in the mass timber market (glulam and mass timber panels) would grow by 800 million board feet annually. The board study did not provide an estimate of mass timber's total softwood lumber demand in 2025. If one interpolates the 2025 incremental estimates back to 2021/2022, however, the estimate of annual lumber usage for 2021 here is reasonable.

The methodology used in the analysis derives estimated lumber consumption from the total square footage of mass timber buildings constructed per year (as publicly reported by WoodWorks); a mass timber usage factor per square foot of building; a lumber usage factor per cubic foot of mass timber; and an adjustment factor to account for mass timber buildings that are not included in the WoodWorks database. The estimate of capacity makes two key assumptions: each North American plant operates at about 65 percent of the maximum press capacity, and about 50 percent of the total capacity in North America is dedicated to making panels for nonbuilding applications.

Finally, the analysis included an adjustment for mass timber panels imported from Central Europe. Anecdotally, mass timber imports from Europe increased substantially over the last several years. However, the mass timber import data as reported by the US International Trade Commission does not show significant import volumes. Thus, the commission data may be underreporting mass timber imports. Future editions of this report will delve into this issue. ◖

---

1   Softwood Lumber Board, *Mass Timber Outlook* (October 2020), https://softwoodlumberboard.org/wp-content/uploads/2021/03/SLB-Mass-Timber-Outlook-2021-Final-Condensed.pdf.

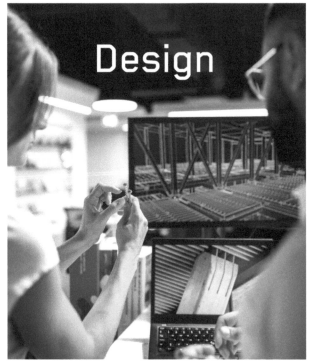

NORDIC
STRUCTURES

nordic.ca

FIGURE 1: THE FORTERRA MODULAR CLT PROTOTYPE 2022 IS AN EXAMPLE OF MASS TIMBER PRODUCTS EVOLVING INTO MASS TIMBER SYSTEMS

Source: Forterra

2022 INTERNATIONAL MASS TIMBER REPORT

Modular and Off-Site Article Outline

MASS TIMBER SYSTEMS: EVOLVING FROM BUILDING PRODUCTS TO PRODUCT PLATFORMS

RYAN E. SMITH, WASHINGTON STATE UNIVERSITY, AND MOD X

IVAN RUPNIK, NORTHEASTERN UNIVERSITY, AND MOD X

TYLER SCHMETTERER AND MOD X

1. INTRODUCTION

The number of mass timber structures continues to grow internationally as the resource, design, testing, manufacture, and assembly supply chains evolve to meet demand. The growing importance of the annual International Mass Timber Conference (IMTC) is evidence of this progress, and it's indicative of the escalating need for networking and knowledge-sharing in the industry. As part of this growth, stakeholders in the mass timber supply chain are evaluating how to increase value from manufacturing building products and structural components, including Cross-Laminated Timber (CLT) and glulam. They aim to strengthen

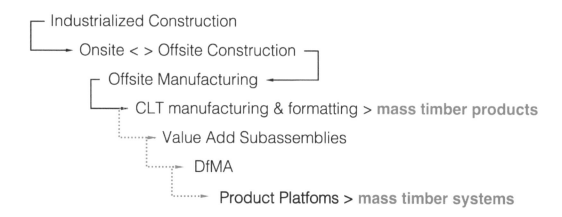

FIGURE 2.1: KEY CONCEPTUAL RELATIONSHIPS IN MASS TIMBER SYSTEMS

Source: MOD X

the value-add of subassembly manufacturing and of integrating mass timber into off-site construction systems (see **Figure 1**). This article addresses the knowledge gap at the intersection of building products and off-site systems by identifying key concepts and explaining how they apply in practice to off-site and modular construction. Selected case studies are cited.

2. KEY CONCEPTS IN MASS TIMBER SYSTEMS

Numerous terms, definitions, and concepts are rapidly being established in the mass timber industry. Mass timber systems are part of an expanding off-site construction practice—the application of manufacturing-oriented processes. This section will identify and define key concepts in manufacturing-based construction and how they relate to mass timber systems (see **Figure 2.1**).

2.1 INDUSTRIALIZED CONSTRUCTION

Industrialized construction is the application of the principles and production methods of industri-

al manufacturing, including advancements in digitalization, automation, lean construction planning and management, data management, and materials science. The goal is to attain increased productivity. These principles can be applied to on-site (e.g., 3D printing) and off-site conditions (e.g., Housing and Urban Development [HUD] manufactured housing). The most successful applications of these principles have resulted in industrializing the off-site and on-site aspects into a continuous whole to increase productivity and optimize labor and materials. "Industrialized construction" is a term widely used internationally. It's also applied specifically in the United Kingdom (UK), where it has become synonymous with modern methods of construction (MMC)—the application of the principles of modern planning and processes. Mass timber construction sometimes utilizes an industrialized or MMC approach to project delivery.

2.2 OFF-SITE CONSTRUCTION

Off-site construction occurs when some portion of a physical building is completed off-site in an enclosed environment. It's often assumed that off-site construction in the US is also industrialized, but there is ample evidence of off-site construction employing small-scale to zero industrial means, methods, and technology. Conversely, some instances of highly industrialized construction approaches are completed on-site (e.g., production housing). Conflating industrialization and off-site construction has resulted in market assumptions that productivity is achieved by moving construction under a roof, an idea that has rarely proven out in practice. On the other hand, highly developed off-site construction companies based internationally perform aspects of construction on-site precisely because they have industrialized the entire process and can accurately calibrate the cost-to-benefit ratio of on-site versus off-site scope-of-work (SOW) decisions. In the US market, the terms "factory-built," "prefabrication," "and off-site manufacturing" are often used interchangeably, with "off-site construction" as the more general term. These designations, however, are manufacturing-specific and represent a distinct step in the value chain of off-site construction project delivery. Off-site construction, therefore, constitutes the entire process of design, planning, manufacture, and site assembly of elements that have been prefabricated off-site. Mass timber products are manufactured in a factory environment, but they are infrequently produced in the service of off-site construction delivery.

2.3 VALUE-ADD AND SUBASSEMBLIES

Although established principles and practices from the manufacturing industry are seldom used in the construction industry, they are crucial when considering industrialized and off-site construction strategies. Within the construction supply chain, traditional on-site construction requires procuring materials and products that are manufactured,

transported to a site, and assembled by trades in piecemeal fashion. As off-site construction continues to gain market penetration, materials and products may undergo value-add in a manufacturing or fabrication stage, during which enhancements are produced to create subassemblies before shipping and site assembly. Subassemblies are the 1D, 2D, and 3D value-add elements (e.g., panels and volumes) that are produced in a factory with enhancements beyond structure alone (i.e., insulation; exterior cladding; interior finishes; mechanical, electrical, and plumbing [MEP], etc.). Subassembly manufacturing may include modifying 1D elements into 2D elements (e.g., open panels) and/or modifying 1D/2D elements into 3D volumetric modules with systems integration. The value-add may be completed by an upstream or downstream supply chain partner, such as a mass timber manufacturer completing the next stage of an enhancement, an existing volumetric modular manufacturer integrating mass timber products into the production process, or a general contractor (GC) setting up a manufacturing operation to self-perform and self-supply mass timber subassemblies for in-house and third-party projects. A new segment of start-up manufacturers is adding value to materials and products by developing subassemblies. Mass timber is starting to integrate value-add enhancements and systems, including 2D panelized and 3D volumetric modules for the off-site construction supply chain (see **Figure 2.2**).

2.4 OPEN VERSUS CLOSED SYSTEMS

Much of the manufacturing of mass timber products, particularly CLT and glulam, consists of the production of bespoke structural components to serve the needs of a specific, one-off project. These components can be conceived of as open systems—systems configured

MASS TIMBER PRODUCTS

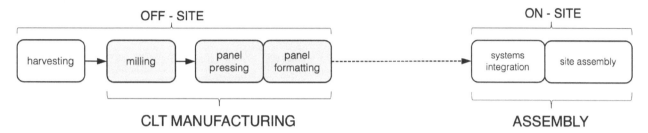

MASS TIMBER SYSTEMS

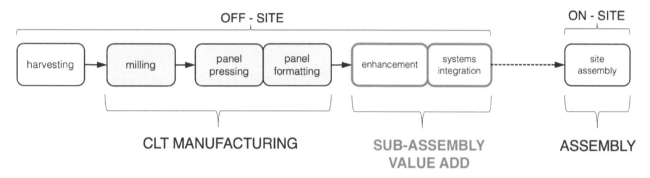

FIGURE 2.2: MASS TIMBER SUPPLY CHAIN RELATIONSHIPS IDENTIFYING OFF-SITE VALUE-ADD MANUFACTURING AND ON-SITE ASSEMBLY

Source: Adapted from T. Beyreuther, Mass Timber Systems

in many ways to produce different buildings. The raw manufacturing of CLT panel blanks, or standard dimensional production output, is an effective illustration of an open-system component. In this context, mass timber is the product of a manufacturing and supply process that is easily integrated into the framework of conventional construction. As the construction industry migrates toward on-site and off-site industrialized methods in planning, manufacturing, and assembly, mass timber components can increasingly function as part of closed systems—systems in which form and structure are two aspects of a single design process. In this manner, mass timber products can be differentiated from mass timber systems in off-site

manufacturing, adding value and specificity to a closed system of construction.

2.5 DESIGN FOR MANUFACTURE AND ASSEMBLY (DFMA)

In the US construction industry, developers, GCs, and some Architecture, Engineering, and Construction (AEC) professionals are practicing a form of closed system design on a project-by-project basis through the application of Design for Manufacture and Assembly (DfMA) principles. DfMA doesn't follow the traditional sequence of project documentation in construction, where means and methods are often finalized late in the process. Instead, DfMA requires early decisions and coordination, often sharing detailed Building Information Modeling (BIM) descriptions among the entire project team to determine the form of manufacturing and assembly that will be used. DfMA is distinguished by the awareness, objectives, and planning of what will be factory-produced versus site-assembled and the labor and workforce skills required to execute the respective work scopes. To gain the maximum benefits of DfMA, subassembly manufacturing of closed systems using mass timber products may be employed to attain some or all of the productivity strategies of industrialization and off-site construction.

2.6 PRODUCT PLATFORMS

The DfMA approach is a significant improvement over the standard model of coordination among project teams and manufacturers, especially when applied to large, unusual, and complex structures. In the context of residential housing, hospitality, commercial offices, industrial structures, and other typologically repeatable structures, a higher degree of design coordination is possible within the framework of industrialized off-site construction via the development of product platforms. An essential organizational tool in all product manufacturing industries, product platforms are defined as a collection of modules or parts that are common to a number of products (or buildings) and developed intentionally to achieve desired effects to create value. While the term "standardization" has historically had negative connotations in AEC communities, building platforms can support a high degree of variability by reusing (part) variants that are interchangeable across multiple configurations to achieve an economy of scale without standardizing the product (or building). In summary, standardization at the part or component scale allows for more variety at the product or building scale.

Product platforms integrate subassemblies and processes that support products designed for manufacture, assembly, and variable configuration within a given framework. This process of predesign is what distinguishes product platforms from project-based DfMA, and it more accurately describes how smartphones and automobiles are designed and manufactured. A product platform may be developed based on a single project, but an investment in customer data input, digital design standards, manufacturing, and construction feedback loops is required to improve the platform so subsequent projects can realize their full potential. This continual improvement cycle is why the approach is often used by vertically integrated companies or joint ventures among stakeholders in the supply chain that collaboratively develop a platform for multiple projects within a consistent group. The approach does not require full vertical integration and offers the inherent value that end users are not required to be familiar with the nuances of a particular manufacturing and assembly process, but work within a well-established framework of developers, developer-builders, or designer-developer-builders.

Product platforms optimize the need to standardize manufacturing and assembly processes with the needs of specific sites and building programs. In this setting, product platforms are the backbone of mass production and bespoke customization or mass customization. While product platforms benefit from vertical integration, particularly during the formative period, they also provide companies the opportunity to separate and outsource aspects of the manufacturing and assembly processes to other subcontractors or suppliers because the parameters

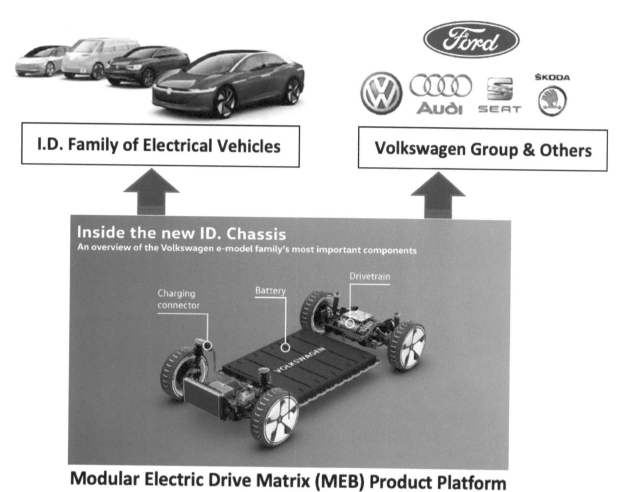

FIGURE 2.3: VOLKSWAGEN ELECTRIC DRIVE MATRIX (MEB) PRODUCT PLATFORM

Source: Volkswagen

of the subassemblies that make up the product platform are clearly specified. Product platforms not only support the synchronous interchangeability of some subassemblies; they also support the continual improvement of subassemblies over time at variable rates from one another.

Volkswagen has embraced this type of manufacturing innovation. The world's second-largest car maker is heavily investing in the future of mobility with a transformative product platform called the modular electric drive matrix (MEB) to supply a modular chassis for electric vehicles (EVs) across several models and brand categories (see **Figure. 2.3**).

By 2028, it will launch almost 70 new e-models to the market while the platform will serve as the foundation for 15 million to 22 million vehicles worldwide. Further demonstrating the potential impact and flexibility of this type of product platform concept, Volkswagen is licensing the technology to competitive automotive manufacturers, including Ford, Fisker, and various EV startups. That way, it can amortize the cost of the platform research and development, decrease barriers to EV market entry, and further accelerate EV market penetration to reduce global carbon dioxide emissions and support corporate sustainability objectives.

In addition to the potential for interchangeability of subassemblies, product platforms can offer this feature to end users throughout the life of the structures. One of the most compelling illustrations of interchangeable flexibility in the world of off-site construction is the post-occupancy variability of Sekisui Heim homes in Japan that offer homeowners kitchen and bath upgrades in addition to disassembly and relocation of entire structural building volumetric modules. Sekisui Heim has been able to support Japan's ambitious national goals of sustainability and a circular economy while also creating a significant secondary market of goods and services to its existing loyal customers by using the inherent principles of their half-century-old and continually improving product platform.[1]

Product platforms are a relatively new concept in the construction industry, accurately describing key conceptual and organizational foundations of manufacturing logics. Without the application of product platforms, no form of construction can realize the productivity potential of industrialized and off-site construction. Further, product platforms are essential to recouping the initial capital-intensive investment of migrating construction off-site, and the acquisition of state-of-the-art mechanized and automated production tools common to mass timber product manufacturing. Conversely, product platforms are a crucial strategy to insulate manufacturers and building stakeholders from the risk of overly productizing building structures (i.e., overstandardization) that cannot adjust to changing market demands, customer preferences, supply chain disruptions, regulatory changes, and future needs.[2] Building on the forthright *Farmer Review of the UK Construction Labour Model*, a 2016 report commissioned by the British government that declare "modernise or die,"[3] the construction industry and the growing mass timber industry will need to "modularize or die" through the development of product platforms to successfully migrate from mass timber as a building product to mass timber as a building system.

3. CASE STUDIES

The case studies table lists several projects that use mass timber products and leverage subassembly manufacturing and off-site construction methods. The subassembly system for each project is identified as a concept, prototype, single project, or multiple projects. Each also includes the subassembly: 2D enhanced panel or 3D volumetric; open versus closed system; and whether it's a product platform (see **Table 3.1**).

Reviewing this collection of case studies, it's evident that mass timber is largely being deployed as a building product. The use of mass timber in projects requires stakeholder teams to accommodate an alternative design and construction approach (and DfMA in some cases), but it does not fundamentally modify project delivery modes. Mass timber is primarily manufactured for one-off conventional construction projects, but it is being considered as part of a value chain in an industrial-

---

1 In 2008, 10 major home providers in Japan introduced a certification system called Sumstock for existing off-site manufactured homes to alleviate homeowner safety concerns including earthquake resistance and potential defects. This effort includes the upgrade of interiors and the entire relocation of volumetric modules. In 2018, MOD X witnessed firsthand the disassembly of a Sekisui Heim house that was being relocated in Tokyo. Additional information is available at *https://japanpropertycentral.com/real-estate-faq/sumstock-certification-for-existing-homes/*.

2 P. Jensen, H. Lidelow, and Olofsson, "Product Configuration in Construction," *International Journal of Mass Customisation, Inderscience Enterprises Ltd.* 5, no. 1 (2015): 73–92.

3 M. Farmer, "Modernise or Die. London, Construction Leadership Council," 2016. Accessed, January 24, 2022. https://www.cs-ic.org/media/2846/mark-farmer-modernise-or-die.pdf.

NUMBER OF PROJECTS	BUILDING USE & SIZE	STAKEHOLDERS	LOCATION(S)	VALUE-ADD SUBASSEMBLY TYPE & MANUFACTURER	OPEN/CLOSED SYSTEM	PRODUCT PLATFORM	REFERENCE LINK
One Project Built (Min.)	3–8 Story Multi-Family	Stora Enso (Fig. 6)	Finland & Global	3D Volumetric	Closed	✓	https://www.storaenso.com/-/media/Documents/Download-center/Documents/Product-brochures/Wood-products/Design-Manual-A4-Modular-element-buildings20161227finalversion-40EN.pdf
Project Built on Stora Enso Platform	Puukuokka Block - Multi-Family Mid-Rise (150 Units)	OOPEAA	Finland	3D Volumetric	Closed	✓	https://oopeaa.com/research/modular-timber-construction/
Multiple	Single Family Detached & Townhouse (2-3 beds)	Legal & General Homes	UK	3D Volumetric	Closed	✓	https://www.landghomes.com/
Multiple	Mid-Rise Multi-Family	Derome AB	Sweden	3D Volumetric	Closed	✓	https://www.derome.se/bostadsutveckling/produkter-tjanster/modulbyggnation
Multiple	Multi-Family Mid-Rise	Sizes Works AB (Manufacturer/GC)	Sweden	3D Volumetric	Closed	✓	https://sizes.se/
Multiple	40+ Multi-Family & Hospitality Projects	Kaufmann Zimmerei und Tischlerei GmbH	Austria	3D Volumetric	Closed	✓	https://www.kaufmannzimmerei.at/holzmodulbau
Multiple	31+ Educational, Hospitality, Multi-Family, & Dormitory	Kaufmann Bausysteme GmbH	Austria	3D Volumetric	Closed	✓	https://kaufmannbausysteme.at/de/raummodule
Project in Development	7 Unit Low-rise Townhouse (7,950 SF)	Green Canopy Node (Developer); Mahlum (Architect); ROC Modular (Manuf)	Washington, USA	3D Volumetric	Open		https://www.greencanopynode.com/construction-technologies
Project Unrealized	312 Unit Mid-Rise Workforce Housing (646,000 SF) *Unrealized Project*	Nexen CNOOC Ltd. – (Developer), Perkins + Will (Architect), ATCO (Manuf)	British Columbia, Canada	3D Volumetric	Open		https://perkinswill.com/
Two Projects Built	Single & Multi-Family Low-Rise	Vestis Systems & Berg Co. (Developer), Vaagen Timbers (Manuf)	Washington, USA	2D Enhanced Panel	Open		https://www.blockhouselife.com/
Multiple	Multi-Family & K-12 Schools	Egoin Wood Group	Spain	2D Enhanced Panel & 3D Volumetric	Open		https://egoin.com/en/
Multiple	Assisted Living, Hospitality, Single & Multi-Family housing	Carbon Dynamic	Scotland, UK	3D Volumetric – Carbon Dynamic	Open		https://carbondynamic.com/
Concept	Modular Prototype for Siingle- & Multi-Familly	Forterra	Washington, USA	3D Volumetric - Zaugg AG Rhrbach	Open		https://forterra.org/subpage/modpro
Multiple	Multi-Family Low-Rise Market Rate & Affordable	Swan Housing with NU Living & various architects	UK	3D Volumetric	Open		https://www.swan.org.uk/
Multiple	Single-Family	Wigo	Latvia	2D Enhanced Panels	Open		https://wigo.info/
Multiple	Multi-Family & Office	URBAN MASS TIMBER – Kunzli Davos AG Timber Construction	Switzerland & Texas, USA	3D Volumetric	Open		https://www.urbanmasstimber.com/
Concept	Mid-Rise & High-Rise Multi-Family	Weber Thompson – Architect; ARUP - Engineering; Sellen - Construction	Washington, USA	3D Volumetric	Open		https://www.weberthompson.com/thought/tall-wood-buildings-have-a-promising-future-with-clt/
Concept	High-Rise Multi-Family	Valerio Dewalt Train Associates, Inc. – Architect; Factory_OS Manufacturer	USA	3D Volumetric	Open		https://www.buildordie.com/vdta-blog/modular-construction
Multiple	Mid-Rise & High-Rise	Intelligent City	British Columbia, Canada	Parametric Digital Platform, Integrated CNC Manfacture	Open		https://intelligent-city.com/platforms-for-life-process/

TABLE 3.1: MASS TIMBER OFF-SITE CONSTRUCTION CASE STUDIES

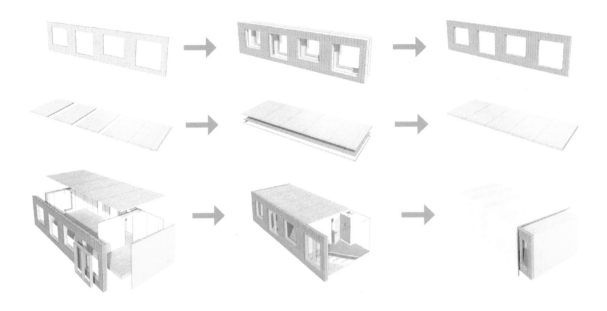

FIGURE 3.1: STORA ENSO MODULAR BUILDING SYSTEMS PROPOSES A CLT-BACKED SUBASSEMBLY FOR VOLUMETRIC MODULAR PRODUCTION THAT CAN BE CONFIGURED AND ASSEMBLED INTO DIFFERENT BUILDING PROJECTS

Source: Stora Enso

ized off-site construction system with subassembly manufacturing enhancements for improved productivity. To leverage the potential of mass timber products, mass timber supply chain stakeholders—developers, architects, engineers, mass timber manufacturers, subassembly manufacturers, and site assemblers/GCs—should consider how to develop mass timber as an integral strategy in the development and delivery of product platforms.

Although several business alignments may achieve positive results, off-site construction and product platform delivery are proving successful in vertically integrated or joint-venture supply chains where the platform investment can be utilized, amortized, and improved across several projects for quality control, cost control, and timeline reduction. As such, the specific supply chain compa-

ny that performs the value-add subassembly manufacturing may vary. It might be an existing supply chain partner such as a CLT manufacturer moving down the supply chain to produce subassemblies or a GC moving up to create value-add elements with mass timber products (e.g., Timberlab). Mass timber subassemblies could also be supplied by an existing panelized or volumetric manufacturer willing to integrate mass timber products into its production line (e.g., ROC Modular). Finally, several start-up subassembly manufacturers and suppliers are focused on value-add, with more expected to follow in this rapidly developing sector.

An illuminative case study focused on developing a clean and openly accessible product platform concept is called *Building Systems by Stora Enso: 3–8 Storey Modular Element Buildings.*[4] Stora

---

4   "Building Systems by Stora Enso: 3–8 Storey Modular Element Buildings," Stora Enso, 2016. Accessed January 24, 2022. https://www.storaenso.com/-/media/Documents/Download-center/Documents/Product-brochures/Wood-products/Design-Manual-A4-Modular-element-buildings20161227finalversion-40EN.pdf.

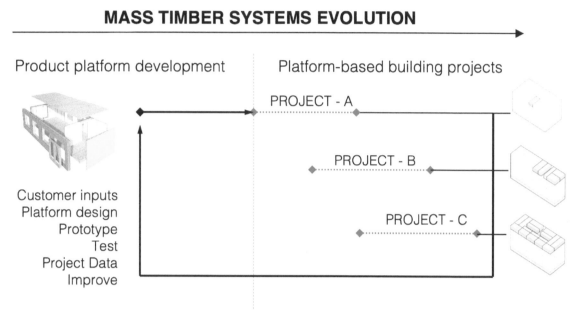

MASS TIMBER SYSTEMS EVOLUTION

Product platform development

Platform-based building projects

PROJECT - A

PROJECT - B

PROJECT - C

Customer inputs
Platform design
Prototype
Test
Project Data
Improve

FIGURE 3.2: PRODUCT PLATFORMS ARE PREDESIGNED WITH CUSTOMER INPUTS AND RESEARCH APPLIED TO MULTIPLE BUILDING PROJECTS WITH A POST-OCCUPANCY FEEDBACK LOOP TO CONTINUOUSLY IMPROVE THE PRODUCT PLATFORM

Source: MOD X

Enso developed a CLT-based 3-story to 8-story residential building platform intended for designers, contractors, building owners, and developers (see **Figure 3.1**). In this system, the subassembly is the CLT-backed floor-, wall-, and roof-enhanced panel elements that are configured into various volumetric modules to be shipped and assembled on-site. The product platform approach of the 2D enhanced panel subassembly integrates into a volumetric modular subassembly and, thereafter, different projects. A single product platform is intended to serve many different projects as a configurable and adaptable system that is continuously improved (see **Figure 3.2**). The 2D subassembly to be standardized with a high degree of flexibility and variability in configuring volumes for customization of a particular project application, demonstrating the evolution of mass timber products in the supply chain from a bespoke structural element assembled on-site to an integral product for the development of a subassembly platform as a building system.

To address this opportunity for mass timber systems, the 2022 IMTC included a new track developed by conference partner MOD X that focuses on off-site and industrialized construction and housing. MOD X will create, coordinate, and moderate a series of four panel discussions as a specialized educational track during the conference and coordinate a local off-site manufacturer tour in April 2022. MOD X will also moderate an interactive panel discussion and virtual factory tour during the IMTC virtual extension day in May 2022. The sessions in the off-site track are designed to address the needs identified in this article for an evaluation of mass timber as product, platform, and system, and they address the following topics: Value-Add Supply Chains; Product Platforms and Modularization in Offsite Construction; Appropriate Digitalization and Automation; and Business Models in Industrialized Housing. ◯

THE MASS TIMBER INTERVIEW

RANDY ABDALLAH,
Executive Vice President, Walbridge

WOJCIECH NIEDZWIEDZ,
Senior Building Services Consultant,
CREE Buildings

A new feature in the *International Mass Timber Report* is an interview with industry leaders. For the 2022 report we reached out to CREE Buildings and Walbridge to discuss their joint efforts in working with mass timber and concrete hybrid designs. We spoke with Randy Abdallah, executive vice president, Walbridge; and Wojciech Niedzwiedz, senior building services consultant at CREE Buildings.

International Mass Timber Report (IMTR): Walbridge has a long history of using traditional construction methods and materials across a wide variety of customers. How did the opportunity to integrate the use of mass timber first get your attention?

Walbridge: The use of mass timber in the construction of buildings has been around for hundreds of years but really gaining popularity in the United States over the past few years. With a keen focus in helping our clients reach their carbon-neutrality goals, Walbridge is continually on the search for sustainable and cost-competitive products that can be incorporated into their projects. We watch for trends in the industry that will add significant value. The use of a modular and sustainable product at a highly competitive

price point is exciting. The CREE solution meets all those requirements.

IMTR: What do you see as the primary benefits of substituting timber for other more traditional construction materials?

Walbridge: The benefits can be numerous and are changing from health and well-being to commercial factors like cost and lead time. As we slowly begin to integrate back to the office and other public gathering spaces, our clients are interested in the biophilic properties timber offers to their valuable human resources. Additionally, as the domestic supply chain for timber expands, we are experiencing faster delivery, and in some cases lower cost on timber products over traditional.

IMTR: Why did you decide to incorporate the timber-hybrid approach?

Walbridge: We really appreciate the flexibility of the timber-hybrid approach; its modular properties allow the right materials to be used at the right time and for a competitive cost. Additionally, it's not a radical change, but one that takes advantage of a supply chain that already exists for Walbridge. It gives our clients a sustainable, financially viable, and available option for the use of mass timber.

IMTR: What are the advantages of the timber-hybrid approach that you are taking?

Walbridge: More than 90 percent of Walbridge's annual revenue is generated by repeat clients; these clients expect the highest standard of safety, quality, schedule, and cost certainty. By design, the CREE system allows hours to be taken out of the field and into a controlled environment to ultimately meet their core expectations.

IMTR: What challenges or obstacles have you had to overcome?

Walbridge: The biggest challenge that we had to overcome was developing our supply chain; although, as I previously mentioned, the partners' capabilities exist, for many this is not business as usual. With no fabricator domestically making the CREE product, we first had to educate and develop our partners before moving forward with the solution.

Some of the other challenges that we have faced revolve around reeducating the public on the benefits of wood. Until recently, our building codes never properly addressed mass timber; and many of the myths still exist around the use of wood, for example, its fire rating and maintainability. However, the industry is making good progress on all fronts.

IMTR: There is a growing number of firms working with timber solutions. How did you decide to work with CREE?

Walbridge: We really believe in CREE's collaborative approach to creating a lasting framework. Rather than pushing a product, they have created a community that drives innovation through sustainable practices.

IMTR: CREE has a history of passionate support for resource-efficient and systemized timber-based construction. How did Walbridge, a traditional general contractor, and CREE meet?

CREE: Our company strategy bringing CREE to the US market meant we had to carefully select our local partners. As a company, we need to make sure we find partners who share a similar company culture and have the intrinsic strategic direction when it comes to driving innovation and changing the way buildings are built. Based on our own search criteria and studying company profiles, we have studied the Engineering News-Record (ENR) Top 100 rankings. Walbridge impressed us in many ways: they are a family business with a market reach that enables us to serve clients in several states; and most importantly, the initial interaction and relationship-building on that has been characterized by trustworthiness and great communication.

IMTR: What do you see as the advantages of the timber-hybrid approach compared to other mass timber projects?

CREE: The construction industry is slow to innovate and to adopt new materials. To convince the planners and developers to change the design from a traditional steel/concrete approach to mass timber can be very challenging and often very expensive. Mass timber structures are usually more expensive because of the greater quantities of wood involved.

The timber-hybrid system nicely marries a common construction material with more price stability, like concrete, with the less common and more price-volatile material like mass timber, without complete reliance on one trade. Therefore, the timber-hybrid system provides a low-risk option in terms of construction capacity and cost, and an efficient solution to address multiple building requirements simultaneously (strength, acoustics, fire, etc.).

It is also important to mention the facts, like better physical performance in terms of vibration, deflection, sound insulation, fire resistance and compartmentalization; weather and water protection during construction and operations; and many more advantages of timber composite against all-timber.

Furthermore, it must be acknowledged that all this comes along with minimized material input, which also secures cost competitiveness, if not even cost parity, against concrete/steel, as proven by some of our recent projects.

Increasing numbers of experts—and clients as well—follow that ideal combination, utilizing the benefits of timber and concrete/steel alike. And this is where we at CREE Buildings see the con-

struction industry heading, featuring any conceivable form of structural timber composite/hybrid.

IMTR: Are there specific types of projects that lend themselves more to the timber-hybrid approach?

CREE: There are two main factors that should be taken into account when selecting a project to suit the timber-hybrid system:

- A systemized approach and integrated project delivery: One of the most important factors is to make sure that the project has a regular grid. The systemized approach of the CREE system allows for more freedom in design, especially in facade and interior architecture. The multiple grid options available for the plan and elevation allow designers to place and arrange fully developed, detailed, and coordinated digital elements. All the system components are interconnected Building Information Modeling (BIM) families with a high level of detail. This lets the architect establish a core and shell project in a fast, efficient, detailed, and analytical way. As a result, the architect has more time and freedom to design a facade and interior spaces that meet the expectations of all the parties involved, as well as complying with environmental and other local regulations. Cost control and time certainty are transparent and fully vetted by all parties involved. This motivates designers and builders to work together more closely and helps them find the most cost-effective design strategy.

- This leads to the second point: integrated project delivery requires all stakeholders to be involved from the very beginning. This changes the usual procedure significantly, as many decisions must be taken much earlier than in the conventional planning process, starting with the selection of the CREE system and its components. However, the benefits quickly become clear: soon afterward, several simulations of all workflows are carried out in the digital twin, which facilitates optimized material and information flow and the elimination of resource wastage. This means that the project delivery process becomes considerably more effective and favors greater efficiency for the whole construction phase, partly through the off-site industrial prefabrication of building components and partly through their assembly on-site. The components are produced according to an environmentally and economically compatible, socially responsible, and sustainable process, as continually developed by CREE.

IMTR: What were the biggest challenges in developing the hybrid approach for this your project?

CREE: As always, it is difficult to convince someone to try something new. Especially when you try to bring a new concept to a different country or even more so to a different continent. This applies in particular to compliance issues with local building codes, fire regulations, design criteria, etc.

We invest lots of time in educating potential customers; developers; general contractors; planners; architects; structural engineers; mechanical, electrical, and plumbing (MEP) specialists; and code consultants. Our ethos is to share the knowledge and experience we have gained over the last 10 years, create a network of like-minded companies and individuals who are willing to collaborate and share their expertise in order to drive the innovation. This has proven to be a successful approach and can be measured in the number of license partners and projects we have delivered all over the world.

By working closely with Walbridge and their supply chain, we are able to transfer our knowledge and to implement it in the projects they are working on. Our marketing and business development department has developed strategies to help our partners to leverage this knowledge and experience and support their tender bids and project acquisition.

By building the first CREE project in Canada and the mock-up building in the US, we believe we will be able to convince even more customers who are new to the concept.

IMTR: How much impact do you think the timber-hybrid solution will have on the rest of the mass timber movement?

CREE: According to our research, 80 percent of the built multistory timber-based buildings (taller than 5 stories) are actually hybrid buildings. They often have a concrete core and, even more often, a concrete podium (ground floor and sometimes first floor too). Other times, they have steel bracing systems integrated in their structures, in their facade, and especially in the US, many projects have their core made of steel. Even projects where the entire structure is made of Cross-Laminated Timber (CLT) (walls, core, floors) are still often designed with a concrete podium. Therefore, the impact of timber's combinations with concrete and steel is already strong, and it represents already the common practice.

It is true that there are some famous projects like Treet in Norway, or Wood Innovation and Design Centre in Canada where all the structural elements are Engineered Wood Products (EWPs), but these buildings represent an exception, and they do not represent the global practice.

In order to do better with the industry as a whole, we need to think about how to penetrate the market in large quantities, suggesting timber construction to every "standard" concrete/steel client, for every "standard" large or tall building project. Architectural icons gain attraction, but "standard" buildings represent the vast majority in numbers and thus can cause by far the most positive environmental impact.

At CREE we understood the importance of using timber together with other materials already 10 years ago, and we are working to constantly optimize the combinations between timber with other materials in order to use them in the best way possible.

Without that smart and material-minimizing mix of timber with concrete/steel, many current buildings would have never seen any timber used at all. For example, the Eunoia Junior College, Singapore's first high-rise college, or the staggering 800,000-square-foot Siemens Campus Module 2 would have finally just been built in concrete/steel. Now they feature the uncompromised beauty of exposed structural wood.

We very much appreciate all-timber buildings. They absolutely deserve their position in modern architecture and certainly can contribute a lot of good for our sustainably built environment. But for many projects, a timber-hybrid approach is a better solution. ◐

BIOS

RANDY ABDALLAH,
Executive Vice President, Walbridge

With over 40 years of construction experience, Randy is the leader of Walbridge's Get Work Team and is responsible for business development, estimating, design management, procurement, and marketing. Randy is a graduate of the Construction Technology program at Fanshawe College in London, Ontario, Canada, and the McCombs School of Business Executive Training Program at the University of Texas.

WALBRIDGE
777 Woodward Ave., Ste. 300
Detroit, MI 48226 USA
www.walbridge.com
messages@walbridge.com

WOJCIECH NIEDZWIEDZ,
Senior Building Services Consultant, CREE Buildings

Wojciech is an experienced MEP engineer, specializing in heating, ventilation, and air conditioning (HVAC) systems and thermal modeling simulations. Before joining CREE Buildings in 2019, he spent over 20 years working with several MEP contractors in the UK. He also coordinates all activities between CREE headquarters and the North American partners and is responsible for the digital content on the CREE platform.

CREE GMBH
Färbergasse 17b
6850 Dornbirn, Austria
www.creebuildings.com
info@creebuildings.com

CHAPTER 1: INTRODUCTION

THE MASS TIMBER EFFECT

- Between 2020 and 2034, the number of mass timber buildings constructed globally will double every two years. The result is that the North American building construction sector will reach carbon neutrality.

Historically, wood's use as a construction material, while extensive, was largely limited to low-rise and light-frame buildings. A light-frame building uses many small and closely spaced members assembled by nailing. Typical light-frame construction features 2-by-4s and 2-by-6s as wall supports, wood joists as floor supports, and rafters as a roof assembly. The application of this light-frame construction style is primarily limited to homes, smaller apartment buildings, and low-rise nonresidential structures.

Now, though, the use of wood in construction is shifting with the game-changing introduction of mass timber in North America. According to Perkins and Will, an architecture and design firm that was an early proponent of mass timber:

"The growing field of mass timber is a fundamental disruption of conventional concrete-and-steel approaches to building design and construction. Instead of limiting wood to low-rise, light-frame applications, we can now reimagine wood as an advanced structural system that produces communities with greater speed, efficiency, and resilience."[1]

This report provides readers with a broad, yet deep, understanding of the North American mass timber industry in 2021. This chapter explains why the report was assembled, defines mass timber, describes how it is used, and introduces the mass timber supply chain concept.

1.1 WHY A MASS TIMBER REPORT?

This report was developed as a companion piece to the International Mass Timber Conference, held annually in Portland, Oregon, beginning in 2016. As evidenced by dramatic year-over-year growth in attendance, the conference has strengthened the mass timber community by providing a forum for the exchange of ideas and information, and for the development of relationships along the supply chain.

Mass timber has captured widespread attention in recent years. Architects, engineers, developers, builders, the forest industry, and community leaders are excited about mass timber's revolutionary potential in building construction. And rightly so.

That's because it's a technology that uses renewable resources, reduces building construction and development costs, increases versatility in building sites, is safe, and yields highly usable structures. It seems that almost every day a new mass timber article or report is released, be it a story on a new mass timber high-rise, the announcement of a new manufacturer, or news about a favorable change in building codes. Information on mass timber is being developed at a phenomenal rate. It can be overwhelming, especially when each new piece of information is specific to just one aspect of the industry. By contrast, this report is intend-

---

1   Sindhu Mahadevan, "Mass Timber: A Primer and Top 5," *Perkins + Will* (blog), November 17, 2017.

ed as a single, comprehensive, in-depth source of North American mass timber information, as it stands in 2021.

It is also important to recognize that the mass timber industry is global. The majority of the annual volume of mass timber panels produced globally is manufactured overseas, mostly in Central Europe. As a result, mass timber building projects are often "exported" to destinations halfway around the globe from the manufacturing plants. Thus, this report includes comments and analyses about the global aspects of the industry.

As the industry continues to evolve, this report will be expanded and updated annually.

1.2   WHAT IS MASS TIMBER?

Mass timber is not just one technology or product. Solid wood (i.e., timber and lumber) has been used as a structural material for millennia. More recently, however, a different class of wood products has emerged. These Engineered Wood Products (EWPs) are a group of construction materials that combine wood's inherent strength with modern engineering. EWPs are manufactured by using adhesives to bind strands, particles, fibers, veneers, or boards of wood to form a composite product. The theory underlying all EWPs is that the process of disassembling wood into small pieces and then gluing them back together results in a product that is significantly stronger than a solid wood product of the same dimensions. In a solid piece of wood, strength-limiting defects such as knots, splits, checks, or decay tend to concentrate in a single area. That defective area is where the wood is most likely to fail. In EWPs, the disassembly and reassembly processes randomize the location of defects and yield products

with predictable strength characteristics. EWPs include structural building materials such as plywood, Oriented Strand Lumber (OSL), Laminated Veneer Lumber (LVL) (see **Figure 1.1**), wooden I-joists, and of particular interest in this report, mass timber products.

1.2.1   MASS TIMBER PRODUCTS

Mass timber products are a distinct class of EWPs. The following sections provide descriptions of the different types of mass timber products developed to date.

Cross-Laminated Timber (CLT)

CLT is a panelized structural EWP that can be used in all major building components (floors, interior and exterior walls, and roofs). It is also used as a ground mat at construction and mining sites, allowing heavy equipment to operate on unstable soils. CLT is made of three or more layers of lumber, each layer oriented perpendicularly to the adjacent layer. (See **Figure 1.2**). The layers are then pressed together with a special adhesive. The lumber is typically preselected, so major defects such as knots and checks are removed prior to lay-up. CLT panels used for building construction are commonly 8 feet to 12 feet wide, 20 feet to 60 feet long, and 3.5 inches to 9 inches thick. Panel length is limited only by press size and highway trucking regulations.

Because the lumber is layered with an alternating grain orientation, the strength, dimensional stability, and fire resistance of CLT panels are significantly greater than for individual boards. CLT is produced in dedicated manufacturing plants with machinery for remanufacturing, finger jointing, and surfacing lumber; glue applicators and

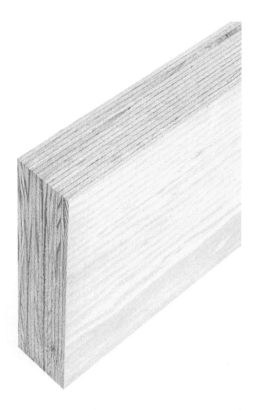

FIGURE 1.1: LAMINATED VENEER LUMBER (LVL)
Source: APA

FIGURE 1.2: CROSS-LAMINATED TIMBER (CLT) PANEL
Source: APA

specialized panel presses; and Computer Numerical Control (CNC) routers that trim panels to size and cut openings for doors, windows, etc.

Most CLT panels are customized for a specific construction project, meaning the exact width, length, thickness (and arrangement of layers), and other properties of each panel are tailored to one building. Openings for doors and windows, as well as openings or channels for electrical, plumbing, and HVAC, are commonly cut by the manufacturer using CNC machines. The prefabricated panels minimize the labor needed at the construction site and dramatically speed construction.

After manufacturing, CLT panels are transported to the construction site, typically by truck. Crews hoist the massive panels into place using cranes, with straps or cables attached to preinstalled "pick points," which are removed once the panel is in place.

In some cases, CLT panels are prefabricated into entire modular units (rooms and/or building sections) that can be transported by truck and installed using cranes, further reducing jobsite construction requirements.

Nail-Laminated Timber (NLT)

NLT is a century-old construction method that recently returned to favor and has been updated with new design guides and construction methods. Like CLT, NLT is a massive wood composite panel. In an NLT panel, however, the wood grain orientation does not alternate. Instead, numerous pieces of lumber are stacked face to face with the

FIGURE 1.3: NAIL-LAMINATED TIMBER (NLT) PANEL

Source: StructureCraft

wide faces adjoining. Rather than using adhesive to bond the layers (as in CLT and glue-laminated timber, or glulam), the lumber is held together with nails (see **Figure 1.3**). Because it does not require the specialized presses used in CLT manufacturing, NLT can be assembled at a temporary workshop close to the construction site or at the building site.

NLT panels are most commonly used in horizontal applications (i.e., floors and roof decks). As a result, fewer precision-machined openings, such as those required for doors and windows, are needed. One drawback is that the metal nails used in NLT can dull or damage woodworking tools such as saws, drills, and routers if the NLT panels are machined. NLT panels can be produced in any thickness common to softwood dimension lumber (e.g., 2-by-4 to 2-by-12). The width and length of the panels are limited only by the application's dimensions. NLT is recognized by the International Building Code (IBC) as being code-compliant for buildings with varying heights, areas, and occupancies.

Dowel-Laminated Timber (DLT)

Dowel-Laminated Timber (DLT) is like NLT, but wooden dowels hold the boards together instead of nails (see **Figure 1.4**). In a process called friction fitting, hardwood dowels are dried to a very low moisture content and placed in holes drilled perpendicularly into softwood boards stacked on edge and side by side. (The wood grain in a DLT panel is parallel.) The hardwood dowels then expand as they gain moisture from the surrounding softwood boards. The result is a tight-fitting connection that holds the boards together. The panel sizes are like CLT and NLT (8 feet to 12 feet wide and up to 60 feet long), but the thickness depends on the width of the softwood boards being used. DLT is most common in floor and roof applications, but StructureCraft, the lone North American manufacturer of DLT, says its panels also can be used in vertical applications.

DLT is the only all-wood mass timber product. With no metal fasteners, DLT panels can be processed with CNC machinery without damage to the cutting tools. That's why DLT is often selected when certain profiles are needed in a panel (e.g., a design to enhance acoustics). The all-wood approach also allows building designers to select a material with no chemical adhesives.

Unlike NLT, which is commonly manufactured at the jobsite, DLT is typically fabricated in a plant, allowing the panels to be manufactured at precise dimensions and to include aesthetically pleasing patterns, integrated acoustic materials, electrical conduits, and other service interfaces.

Dowel-Bonded CLT

Dowel-bonded CLT is a massive, prefabricated, cross-laminated panel with layers of rough-sawn

FIGURE: 1.4 DOWEL-LAMINATED TIMBER (DLT) PANEL
Source: StructureCraft

boards that are bonded with hardwood dowels. This is the newest of the CLT products and should not be confused with DLT, described above. The low moisture content and tight fitting of the dowels at the time of assembly assures a durable tight connection once the dowels swell after gaining moisture in ambient conditions. The panels are assembled in highly automated lines. Only two commercially successful systems are known to date: one developed by Thoma Holz100 (or Wood 100) in Austria and another developed by Swiss industrial hardware manufacturer TechnoWood AG. By mid-2019, TechnoWood had installed 8 highly automated lines in Europe. Unlike other CLT products, some layers of the dowel-bonded CLT are arranged at 45 degrees or 60 degrees to the surface layer direction.

Nail-Bonded Solid-Wood-Wall or Massiv-Holz-Mauer

Massiv-Holz-Mauer (MHM) is a massive, prefabricated cross-laminated panel with layers made of rough-sawn boards bonded with nails. This product should not be confused with NLT that is produced in North America. The nail-bonded MHM (which literally means "mass-wood-wall") technology might have predated the development of adhesive-bonded CLT, but the real breakthrough came with a solid wooden wall system patented in Germany in 2005. MHM is fabricated on small-scale, turnkey, three-step Hundegger production lines. Panels may consist of 9, 11, 13, or 15 layers (each about 16.5 millimeters, or $^{10}/_{16}$ inch). The intended use of this product is as load-bearing and division walls for low-rise buildings with moderate exposure to moisture (below 20 percent) and low to moderate exposure to corrosion.

Mass Plywood Panel

A Mass Plywood Panel (MPP) is another innovative mass timber product that's produced at a single plant located in Oregon, Freres Lumber Co. MPPs are veneer-based, rather than lumber-based, and are constructed by gluing together

FIGURE 1.5: MASS PLYWOOD PANEL (MPP)

Source: Oregon Department of Forestry

many layers of thin veneer in various combinations of grain orientation (see **Figure 1.5**). The uses of MPPs are similar to those of other mass timber panels, though the manufacturer says veneer-based panels can form thinner panels and/or longer unsupported spans than are possible with lumber-based panels.

Glulam

Glulam is another form of mass timber, an engineered wood composite made from multiple layers of lumber. The grain is oriented in a parallel direction in all layers, and the layers are bonded with adhesive to form a structural element with large dimensions. (See **Figure 1.6**) Glulam, a well-established product that has been used in range from 6 inches to 72 inches, and lengths can surpass 100 feet.

Glulam beams are typically much stronger than an equivalent-size solid sawn beam and can be manufactured in customizable sizes and shapes, including cambered, curved, or arched structures. If glulam is to be used in applications where both structural support and appearance are considerations, it is available in four appearance grades: framing, industrial, architectural, and premium.

Post and Beam

Post and beam is a construction method that uses large timbers in both vertical and horizontal applications to create the building framework (see **Figure 1.7**). It allows for large open spaces within the building and flexible wall structuring. Construction using large-dimension (6 inches thick and larger) lumber has been popular in high-end homes for years, but it is now enjoying increased popularity in a variety of larger nonresidential and multifamily residential buildings (office buildings, schools, and warehouses). In these larger buildings, structural loads are typically higher than for single-family residences, so larger-dimension posts and beams and/or engineered wood composites such as glulam are of use. In many cases, post and beam frames make up the structural element of a building frame, while nonstructural walls are commonly constructed with light wood framing.

In structures where mass timber panels are used for the floor, wooden posts and beams are often used for the supporting vertical structural elements.

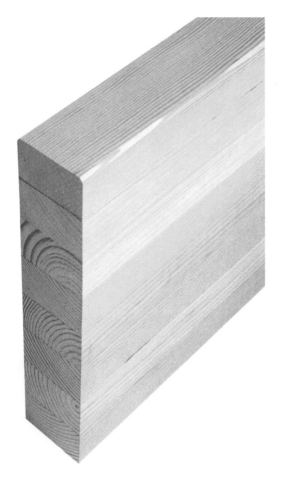

FIGURE 1.6: GLULAM TIMBERS
Source: APA

residential and nonresidential construction for many years, is typically used as either a beam in a horizontal application or as a column in a vertical application because of its high strength-to-weight ratio. Glulam use is on the rise because it is commonly used to support mass timber panels. Other less common uses are as members of massive truss systems or as large-scale utility poles.

Most glulam is made from standard-dimension lumber (e.g., 2-by-4s to 2-by-12s). Thus, typical widths range from about 2.5 inches to 10.75 inches. The potential thicknesses and lengths of glulam, however, are much larger. Glulam depths

FIGURE 1.7: POST AND BEAM

Heavy Timber Decking or Jointed Timbers

Heavy timber decking is used in horizontal applications (floors and roofs) where the full engineered properties of panelized products such as CLT are not required. Heavy timber decking consists of a single layer of timbers (usually 3-by-6s or 4-by-6s) joined edgewise with tongue-and-groove profiles on each piece that lock them together (see **Figure 1.8**). The pieces may be solid-sawn or glulam. Timber decking is more frequently used in regions where construction labor is less expensive, giving this labor-intensive application a cost advantage over other mass timber panels.

FIGURE 1.8: HEAVY TIBER DECKING
Photo Source: Southern Wood Specialties

Mass Timber Hollow Core Panels and Mass Timber Ribbed Panel Assemblies

Mass Timber Hollow Core Panels use thinner (3-ply) CLT panels for the top and bottom layers that are connected with internal glulam ribs. The

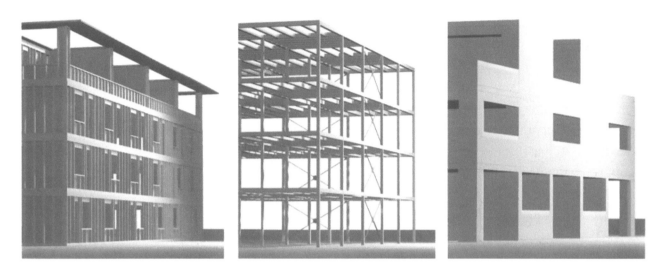

FIGURE 1.9: WOOD-BASE BUILDING CONSTRUCTION SYSTEMS

Source: Fast and Epp

hollow spaces are filled with insulation materials. Mass Timber Ribbed Panel Assemblies are another relatively new mass timber product, combining CLT decks with integrated glulam ribs connected by screws, glues, or a combination of both on the bottom. Both products are typically used as horizontal elements, (e.g., high-capacity floors with extended spans).

1.3 HOW IS MASS TIMBER USED?

Figure 1.9 illustrates how mass timber construction differs from more traditional wood construction.

Light wood-frame construction (building on left) is the most familiar construction system. At a given site, a building is constructed using light wood materials. For example, studs form vertical wall members; joists are the horizontal floor supports; rafters provide roof supports; and plywood or OSB panels sheathe the walls, floors, and roof.

This style is most commonly used in single-family homes and multifamily low-rise housing.

Post and beam construction (center building) involves the use of large, heavy timbers in either sawn or roundwood form. The timbers used as horizontal beams in this style of construction transfer structural loads to other timbers aligned vertically. Diagonal braces between the horizontal and vertical elements provide even more rigidity to the structure. This style allows for an open design because all load-bearing members are fixed points instead of an entire wall.

Mass timber panel construction (building on the right) involves the use of large, solid wood panels for the roof, floor, and walls. Mass timber allows for the construction of wooden buildings that are much taller than light wood-frame construction. There are many forms of mass timber panels, including CLT, NLT, DLT, and MPP. The term "mass timber" as used in this report refers to all of the preceding forms.

1.4 DEFINING THE MASS TIMBER SUPPLY CHAIN

A mass timber supply chain is rapidly developing in North America, and examining the components of that supply chain offers a way to organize and think about this industry. It is important to note that most mass timber products are not standardized commodities. Rather, the fabrication of mass timber products is perhaps best thought of as a step in an integrated process of producing a finished building—the actual final product. Accordingly, the supply and value chains of the mass timber panel industry represent an integrated combination of what is typical for manufacturers of structural EWPs and for the design-engineering-construction sector.

The supply chain starts with the forest resource and flows all the way through to the occupants of a mass timber building (see **Figure 1.10**). In this report, we assess the state of each link in the supply chain, and we address issues such as sustainability, economics, and technology. In short, this report analyzes how people and policies impact mass timber and what that might mean for its development.

1.5 MEASUREMENTS AND CONVERSION FACTORS

Wood products—including logs, lumber, and mass timber products—can be measured and labeled in a variety of ways, some of which can be confusing to those not familiar with common industry practices. This section discusses the terminology, measurement, and conversion conventions used in this report.

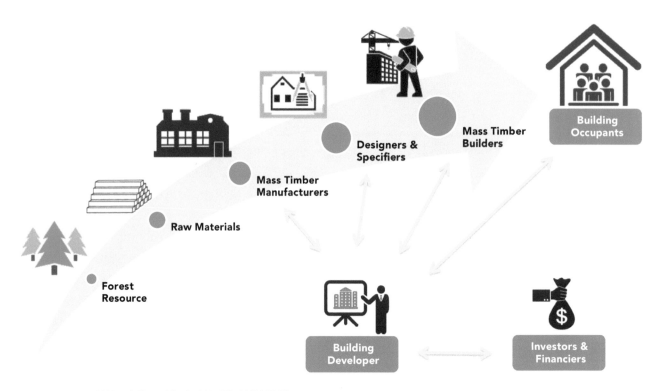

FIGURE 1.10 MASS TIMBER SUPPLY CHAIN

1.5.1 LOG MEASUREMENT

Standing timber and log volume is reported on a cubic foot basis. Cubic feet can be converted to cubic meters using the standard conversion of 35.315 cubic feet per cubic meter. In contrast, a variety of measurement units are used when logs are sold, especially in the United States. In fact, different measurement systems are used regionally, including a variety of log scales and weight-based measurements. Analysis of these systems is beyond the scope of this report.

1.5.2 LUMBER MEASUREMENT

In mass timber, two main types of solid sawn lumber (not engineered wood or wood/glue composite) are relevant. The first is dimension lumber (most commonly 2 inches thick and 4 inches to 12 inches wide). When used in mass timber panels, multiple pieces of dimension lumber are fastened or glued together to create one larger mass of wood. Mass timber is bought and sold in cubic feet (or cubic meters). Dimension lumber is bought and sold in board feet. A board foot is defined as 1-by-12-by 12. Thus, a cubic foot of wood contains 12 board feet. However, a peculiarity of dimension lumber is that its volume is expressed as a nominal size that is larger than the actual finished size. For example, a 2-by-4's actual dimensions are really 1.5 by 3.5. This difference in dimension lumber's nominal and actual sizes means that a cubic foot of wood in a mass timber panel requires an estimated 22.5 board feet of lumber, after accounting for the differences between nominal and actual sizes and yield loss when converting lumber into mass timber panels.

Glulams • Timber Framing
Cross Laminated Timbers

TFC TIMBER FRAME CONNECTION

www.timberframeconnection.com

NOMINAL SIZE				ACTUAL (DRY, SURFACED) SIZE				Conversion Factor (CF/BF)	Conversion Factor (BF/CF)
Thickness (inches)	Width (inches)	Length (feet)	Volume (board feet)	Thickness (inches)	Width (inches)	Length (feet)	Volume (cubic feet)		
2	4	20	13.33	1.5	3.50	20	0.73	0.055	18.3
2	6	20	20.00	1.5	5.50	20	1.15	0.057	17.5
2	8	20	26.67	1.5	7.25	20	1.51	0.057	17.7
2	10	20	33.33	1.5	9.25	20	1.93	0.058	17.3
2	12	20	40.00	1.5	11.25	20	2.34	0.059	17.1

TABLE 1.1: NOMINAL DIMENSION LUMBER SIZES VS. ACTUAL CUBIC MEASUREMENT

Table 1.1 compares the board feet per piece based on nominal size with the actual cubic volume per piece of dry, surfaced framing lumber sold in North America. For consistency, 20-foot-long pieces are used for all examples. The resulting conversion factors (board feet per cubic foot and vice versa) are shown in the two columns on the right side of the table.

The second type of solid sawn lumber used in mass timber structures is heavy timber; it is used as structural support for mass timber panel systems. Heavy timbers may either be sawn to sizes similar to nominal-dimension lumber sizes ("standard sawn") or to the full stated size ("full sawn"). Most heavy timbers are custom-made where the buyer and seller agree on the specified sawn dimensions. For timbers that are full sawn, the appropriate conversion would be 12 board feet per cubic foot.

Globally, lumber practices can vary. In contrast to the North American market, structural lumber in Europe and in many other regions is offered in a variety of thicknesses. Although all lumber imported to the US from overseas conforms with US standards, mass timber panels produced overseas will freely incorporate layers of various thicknesses to meet the required engineering specifications with better efficiency.

1.5.3 LOG TO LUMBER VOLUMES

In the sawmill industry, lumber yield—the volume of lumber produced from a given volume of logs—is expressed in a variety of ways, with regional differences based on local conventions for measuring logs. A full description of these various lumber yield measurements is beyond the scope of this report. But to understand how lumber volumes relate to log demand and harvest, it is most useful to consider cubic yields.

Cubic lumber yields at sawmills vary depending on several factors, with the most important being the log size (diameter). In North America, typical cubic lumber yields for sawmills producing dimension lumber are in the range of 35 percent to 60 percent, meaning that 35 percent to 60 percent of the log volume comes out as finished (dry, surfaced) lumber and the balance is a by-product (chips, sawdust, and shavings), with some volume lost to drying shrinkage. The regions with the largest logs (9 inches to 11 inches average bucked sawmill-length log diameter in the US West) achieve higher cubic lumber yields, while those with the smallest logs (4.5 inches to 6 inches average bucked log diameter in eastern Canada) are on the lower end of the range.

For a quick but rough conversion, multiply a known lumber volume by 2 to estimate the log volume required. For example, to produce 100 cubic feet of dimension lumber, a mill needs 200 cubic feet of logs.

1.5.4 MASS TIMBER PANELS AND GLULAM

Most measurements of mass timber panels and glulam beams are expressed in terms of cubic feet or cubic meters. These figures are based on the actual size of the finished product (although cutouts and channels are typically not deducted). For example, a CLT panel that is 6 inches thick by 10 feet wide and 40 feet long would measure 200 cubic feet ([6 ÷ 12] x 10 x 40), or 5.66 cubic meters (200 ÷ 35.315).

When considering the amount of lumber used in mass timber or glulam products, it is important to consider the nominal vs. the cubic size of the lumber feedstock (see **Table 1.1**), as well as any volume lost during the manufacturing process. In CLT, DLT, and glulam, the lumber is surfaced during the manufacturing process, with about

WOOD VOLUME	VOLUME OR CONVERSION FACTOR	UNIT	DESCRIPTION
Mass Timber Volume	100,000	Cubic Feet	Total CLT and glulam used in building project
	22.5	BF per CF	CLT/glulam to nominal lumber conversion
Dimension Lumber Volume	2,250,000	Board Feet	Purchased dimension lumber
	0.057	CF per BF	Conversion from nominal to cubic volume
Cubic Lumber Volume	128,250	Cubic Feet	Equivalent cubic volume of lumber used
	0.5	CF per CF	Cubic lumber yield from logs
Log Volume	256,500	Cubic Feet	Log demand from mass timber project

TABLE 1.2: SUPPLY CHAIN CONVERSIONS EXAMPLE

$1/16$ of an inch removed from all four sides (exact amounts vary by manufacturer). Also, some volume is lost when defects are trimmed from lumber feedstock, and when panels or beams are trimmed to final dimensions.

For typical CLT or glulam manufacturing, a total of 20 to 25 nominal board feet of dimension lumber is used per cubic foot of finished product.

1.5.5 MASS TIMBER TO LOGS EXAMPLE

Given all the preceding measurement and conversion conventions, it is possible to approximate the total amount of timber (logs) required for a mass timber project. **Table 1.2** follows the wood back through the supply chain to estimate the total lumber and then the logs required for a hypothetical building project that uses 100,000 cubic feet of CLT and glulam. This calculation is only an estimate, and it depends on several assumptions (lumber yield, size of lumber used, and CLT and glulam wood use), but it provides a reasonable indication of the wood volume at various points in the supply chain.

The results show that substantially more log volume is required than will be reflected in the finished product volume. Importantly, the material not used in the final mass timber product is not wasted. Depending on the region where the lumber and mass timber are manufactured, the by-products can be used in a variety of ways. Chips are typically used for making paper. Sawdust or planer shavings make composite panels (particleboard or medium-density fiberboard). By-products can also be made into wood pellets for heating or power generation, or combusted in a boiler to generate power and/or provide thermal energy for lumber drying or other uses.

CHAPTER 2: THE FOREST RESOURCE

IMPACTS OF THE MASS TIMBER EFFECT

- Forestland area in the United States and Canada has been stable for more than 100 years. A contributing factor to that trend is that making products from trees, like mass timber, creates an incentive for landowners to maintain their lands as forest.

- Both US and Canadian forests have the capacity to sustainably supply more timber as the market for mass timber buildings grows. We expect this to contribute to maintaining and possibly increasing the area of North American forests in the future.

- US and Canadian forests can simultaneously provide wood products via harvesting and supply other ecological functions, such as clean air, clean water, carbon storage, recreation, and wildlife habitats.

- This chapter focuses on forests—the beginning of the mass timber supply chain. In the first section, the area, ownership, and types of forests in North America are analyzed. In the second section, the focus shifts to how those forests are managed, including their ability to sequester and store carbon relative to their ability to sustainably provide raw materials for mass timber construction and many other applications.

2.1 CHARACTERIZING THE NORTH AMERICAN FOREST RESOURCE

Forests are a key component of the landscape in many regions. Accordingly, **Figure 2.1** illustrates the portions of North America with more than 15 percent tree cover. As shown by the color shadings that represent different types, forests are made up of coniferous (softwood) trees in the coastal and moun-

tainous areas of the West; mixed hardwood and coniferous trees in the US Midwest, Eastern US, and Canada; and coniferous trees in the US Southeast. Also note that the far northern regions of Canada and Alaska have vast areas of boreal forests. But, given their distance from major population centers and their smaller tree sizes, these forests have little commercial value for conversion to lumber. Finally, in Central Europe, Germany and Austria are leading production of high-performing mass timber elements, including glulam and Laminated Veneer Lumber (LVL) made of local hardwoods, mostly beech and oak. These technologies are commercialized, and a similar development may be expected in North America, with a focus on hardwood species abundant in the US Other region.

As further discussed in Chapter 3, lumber-producing regions are often defined by the types of softwood they commonly produce. The five main regions in North America are the US West, US South, US Other, Eastern Canada, and Western Canada. Forests in the US West are dominated by Douglas fir, Western hemlock, and various pine species. In both Eastern and Western Canada, forests are largely composed of various mixtures of spruce, pine, and fir (SPF). In the US South, four types of pine—loblolly, slash, shortleaf, and longleaf—are the leading species of note for mass timber. When sold as lumber, those four species are lumped into a group called Southern Yellow Pine (SYP). The US Other region includes the Upper Midwest and Northeast. These forests are more heavily stocked with hardwood trees and therefore up to this point have been less significant from a mass timber industry perspective. However, in Central Europe, Germany and Austria are leading production of high-performing mass timber elements (glulam,

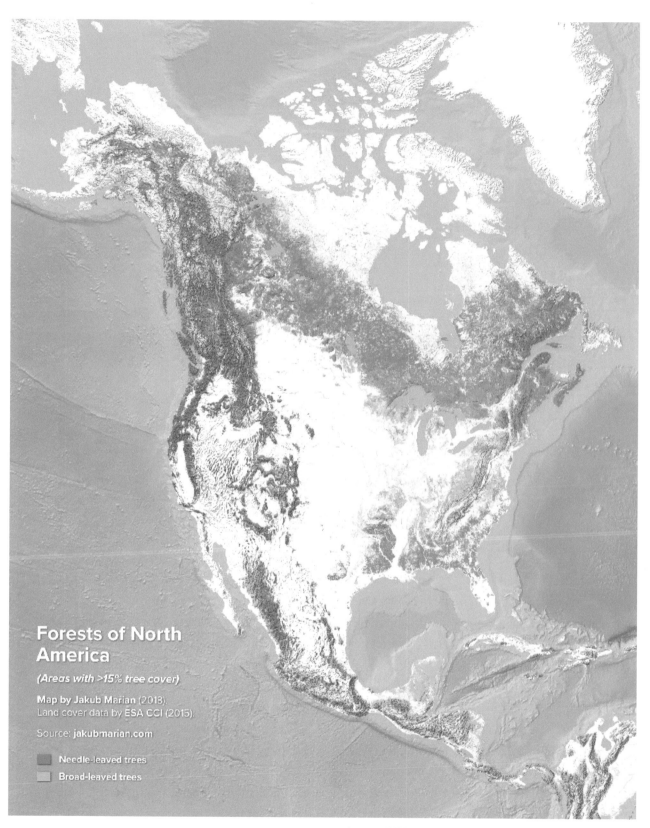

Forests of North America

(Areas with >15% tree cover)

Map by Jakub Marian (2018).
Land cover data by ESA CCI (2015).

Source: jakubmarian.com

Needle-leaved trees
Broad-leaved trees

FIGURE 2.1: EXTENT OF FORESTS IN NORTH AMERICA

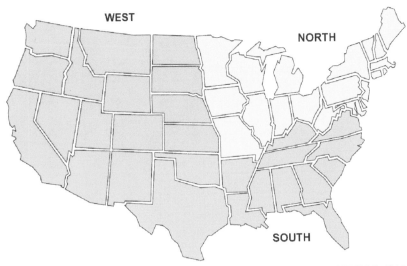

FIGURE 2.2: MAP OF US FOREST REGIONS

FOREST TYPE	NORTH	SOUTH	WEST	TOTAL
Timberland	164,894	208,092	141,437	514,423
Reserved	9,447	5,827	65,290	80,564
Other/Woodland	1,448	54,114	171,846	227,408
Total	175,789	268,033	378,573	822,395

TABLE 2.1: EXTENT OF FORESTS IN THE UNITED STATES BY TYPE & REGION (ACRES IN 1000S)

LVL) produced from local hardwoods (mostly beech and oak). Technologies utilizing hardwood are commercialized in Europe, and similar development may be expected in North America using species common in the US Other region.

Extent of US Forests

The total US land area is about 2.3 billion acres. As illustrated in **Table 2.1,** forests in the United States total about 822 million acres, or roughly one-third of the US land area. Note that this data is from

Forest Resources of the United States, 2017,[1] an update to the 2012 version used in prior versions of this report. The total forest area increased from 766 million acres to 822 million acres in the most recent assessment. The area of forested land in the US has been stable (or increasing, per the most recent analysis) since the early 1900s, despite the US population tripling during the same period. Despite the massive growth in population and the associated increase in demand for wood fiber, it's encouraging to consider that the forest area in the US has remained stable for more than 100 years.

---

1   Sonja Oswalt, W. Brad Smith, Patrick D. Miles, and Scott Pugh, *Forest Resources of the United States,* 2017 (2019), https://www.fs.fed.us/research/publications/gtr/gtr_wo97.pdf.

REGION	NATIONAL FOREST	OTHER PUBLIC	PRIVATE CORPORATE	PRIVATE NONCORPORATE	TOTAL
North	10,147	26,852	30,196	97,700	164,895
South	12,258	13,699	63,504	118,632	208,093
West	73,733	18,584	23,455	25,665	141,437
Total	96,138	59,135	117,155	241,997	514,425

TABLE 2.2: OWNERSHIP OF US FORESTS DESIGNATED AS TIMBERLAND BY REGION AND OWNER TYPE (ACRES IN 1000S)

YEAR	SAWTIMBER	POLETIMBER	SEEDLING/ SAPLING	NONSTOCKED	TOTAL
1953	201,491	170,688	94,565	42,110	508,854
1977	223,210	136,694	115,842	16,607	492,353
1987	242,864	137,981	97,413	8,057	486,315
1997	258,680	127,169	110,283	7,533	503,665
2007	280,265	128,896	96,177	8,875	514,213
2012	294,964	123,144	93,140	9,906	521,154
2017	299,716	117,637	87,395	9,676	514,424

TABLE 2.3: HISTORY OF TIMBERLAND AREA IN THE US BY STANDING SIZE CLASS (ACRES IN 1000S)

The broad category of forested land includes several subcategories: timberland, or forests that are well stocked and capable of producing at least 20 cubic feet of wood fiber per acre per year; reserved forestland, or forests where harvesting of trees is prohibited, mainly wilderness areas and national parks; and woodland/other, where tree cover ranges between 5 percent and 10 percent, tree growth is marginal, and timber production is not a priority. **Figure 2.2** shows the location of the regions listed as columns in **Table 2.1.**

Ownership of US Forests

Timberland is the most productive forest acreage in the US. **Table 2.2** categorizes it by two types of public owners and two types of private owners. As the data in the table shows, higher percentages of timberland are in private ownership in the North and South than in the West. Ownership is important because it affects how land is managed. In general, corporate timberlands are managed to maximize timber production, while public and noncorporate private lands are managed for a broader set of objectives.

Table 2.3 shows a history of the area of US forest designated as timberland classified by tree size, including sawtimber, poletimber, seedling/sapling, and nonstocked. Note that sawtimber includes trees big enough to be sawed into lumber; pole trees are too small for use as sawlogs; seedling/

REGION	SOFTWOOD	HARDWOOD	TOTAL
North	68,278	245,926	314,204
South	149,800	227,981	377,781
West	380,794	43,232	424,026
Total	598,872	517,139	1,116,011

TABLE 2.4: US STANDING TIMBER INVENTORY ON TIMBERLAND BY REGION AND SPECIES GROUP (CUBIC FEET IN MILLIONS)

REGION	1953	1977	1987	1997	2007	2017
North	27,053	43,850	47,618	49,374	55,864	60,601
South	60,462	101,208	105,613	104,846	118,472	141,307
West	344,279	321,902	314,344	329,622	357,264	358,617
Total	431,794	466,960	467,575	483,842	531,600	560,525

TABLE 2.5: HISTORY OF US SOFTWOOD STANDING TIMBER INVENTORY ON TIMBERLAND BY REGION (CUBIC FEET IN MILLIONS)

sapling are very young stands; and nonstocked is bare land that typically has yet to be replanted just after harvest. As the data shows, the area of sawtimber that could be used to make lumber for mass timber has increased by nearly 100 million acres over the last 65 years. This is an encouraging finding.

US Standing Timber Inventory

The US Forest Service is a federal agency charged with managing nearly 190 million acres of national forests and grasslands. Its Forest Inventory and Analysis (FIA) program was established nearly 100 years ago to monitor the conditions of all the nation's forests. A key accomplishment of the FIA was to establish more than 325,000 permanent growth plots in US forests. Each plot is revisited regularly, and data is collected about the trees within the plots' boundaries. Through this system, the FIA can track changes in the forests' statuses. The FIA tracks, for example, key metrics such as species, diameter, age, and cubic volume as part of its inventory of standing trees.

Table 2.4 shows the most recently available (2019) estimate of standing timber volume in the US on timberland acres. As shown, the US has an estimated 1.1 trillion cubic feet of standing timber. The standing volume is relatively evenly split between hardwoods and softwoods.

More specific to the mass timber industry is Table 2.5. It shows the history of softwood standing timber inventory by region. Note that over roughly the last 65 years, the total volume of standing timber in the US has increased by nearly 30 percent in total and by more than 230 percent in the South, both positive findings given the anticipated increased demand from the mass timber industry.

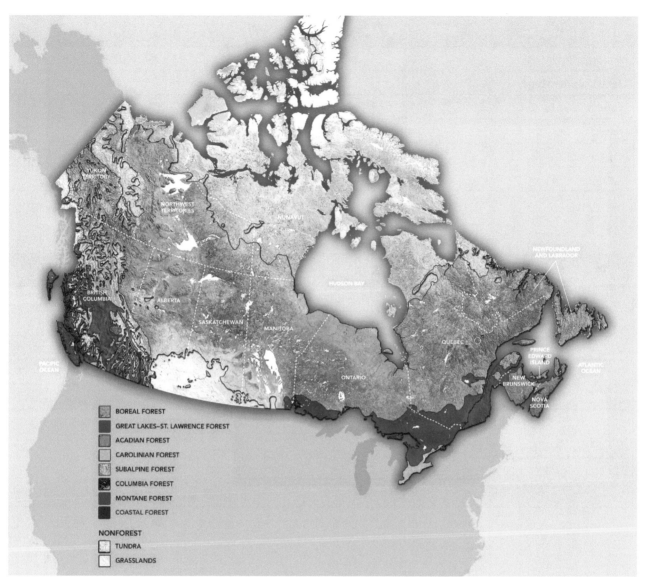

FIGURE 2.3: CANADIAN FOREST REGIONS

Source: Natural Resources Canada. Accessed at https://www.nrcan.gc.ca/sites/nr-can/files/forest/SFM/classification/Canada_forest_regions.pdf

2.1.1 CANADIAN FORESTS

The Extent of Canada's Forests

Canada's total land area is about 2.467 billion acres. Of that total, about 857 million acres is forested. Canada and the US have roughly the same total land areas and total forested areas. Canada's forest area has been stable for decades. **Figure 2.3** shows that there are several distinct types of forest in Canada. The largest and most commer-cially important types include a vast boreal forest that stretches the length of the country from east to west and is composed mainly of spruces, firs, and, to a lesser extent, pines; the forests around the Great Lakes, which are primarily hardwoods, including maple and birch; the montane forests of Western Canada, which are populated with Douglas fir, hemlock, and pines; and the coastal forests in Western Canada, which are heavy with cedar, hemlock, and firs.

OWNER TYPE	PERCENT OWNED
Provincial Crown Land	77%
Territorial Crown Land	13%
Federal Crown Land	2%
Private	6%
Indigenous	2%
Total	100%

TABLE 2.6: OWNERSHIP OF CANADIAN FORESTS

YEAR	CUBIC FEET (IN MILLIONS)
1990	1,684,796
1995	1,682,783
2000	1,671,058
2005	1,623,808
2010	1,607,034
2015	1,594,356
2016	1,591,567
2017	1,585,493
2018	1,575,640

TABLE 2.7: HISTORICAL TOTAL STANDING TIMBER VOLUME IN CANADA (CUBIC FEET IN MILLIONS)

Ownership of Canadian Forests

Over 90 percent of Canadian forests are publicly owned. **Table 2.6** shows a categorization by owner type that includes provincial Crown land, territorial Crown land, federal Crown land, private, and indigenous.

Canada's Standing Timber Inventory

The standing timber inventory in Canada as of 2018 was 1.575 trillion cubic feet, approximately one-third more standing timber volume than the United States, according to *The State of Canada's Forests 2020*.[2] **Table 2.7** shows that Canada's standing timber volume has declined since 1990. The causes of this are many, but two key factors are extensive insect outbreaks and wildfires.

Table 2.8 provides a more detailed estimate of standing timber volume, with categorizations by forest type and stand age. As the data shows,

2   *The State of Canada's Forests 2020*, https://www.nrcan.gc.ca/our-natural-resources/forests-forestry/state-canadas-forests-report/16496.

SPECIES GROUP	1 TO 20	21 TO 40	41 TO 60	61 TO 80	81 TO 100	101 TO 120	121 TO 140	141 TO 160	161 TO 180	181 TO 200	201+	TOTAL
Coniferous	8,665	31,452	70,477	139,785	312,436	224,080	105,282	50,915	33,388	24,018	205,958	1,206,457
Mixed	9,510	17,712	47,769	64,457	83,107	21,329	9,663	4,354	1,488	470	884	260,743
Broadleaf	4,857	11,009	53,038	65,236	44,725	16,482	5,892	1,856	536	33	162	203,827
Total	23,035	60,181	171,294	269,482	440,289	261,893	120,837	57,126	35,414	24,521	207,003	1,671,075

TABLE 2.8: CANADIAN STANDING TIMBER VOLUME BY SPECIES GROUP AND STAND AGE CLASS (CUBIC FEET IN MILLIONS)

more than 70 percent of Canada's forests are co-niferous (i.e., softwoods).

2.2 FOREST SUSTAINABILITY

People across the globe are interested in access to clean air and water. Forests are key to providing access to both. Thus, assuring forest sustainabil-ity is critical to all global citizens. Sustainability is defined as meeting current needs via the con-sumption of natural resources without jeopardiz-ing the ability of future generations to meet their needs for the same natural resources.

2.2.1 GROWTH TO DRAIN

One measure foresters use to monitor sustainabil-ity is "Growth to Drain." Growth to Drain is a ratio of the amount of wood fiber a given area can grow annually (net of natural mortality from insects, disease, fire, etc.) to the amount of wood fiber harvested annually. A ratio greater than 1 indicates that the area is adding more wood fi-ber each year through net growth than is being removed by harvesting. Although many other considerations relate to sustainability, Growth to Drain is frequently a key consideration in forest management and timber harvest planning. The following sections provide an analysis of Growth to Drain for US and Canadian forests.

US Timberlands Growth to Drain

As long as the ratio of Growth to Drain is great-er than 1, forests can supply fiber in perpetuity. **Table 2.9** provides information about historical Growth to Drain ratios in the United States. At the top of the table is data for all softwoods in the US; in the middle is information about hard-woods; and, at the bottom, softwoods and hard-woods are combined. As the data indicates, the ratio is greater than 1 in all cases.

This is a positive finding for the mass timber in-dustry because it indicates that US forests are not being overharvested. However, the data shows a troubling trend. Natural mortality—trees dying from causes such as wildfire, drought, insects, and disease—increased by 250 percent from 1976 to 2016. There is considerable debate about whether the cause is climate change or lack of management, especially in publicly owned forests in the US West. In any case, pressure on Growth to Drain ratios would ease considerably if more trees were used through harvesting rather than lost to natural mortality.

	1976	1996	2006	2016
Softwoods: Annual Mortality (ft³ in 1000s)	2,466,137	3,959,580	4,510,607	5,899,508
Softwoods: Annual Harvest (ft³ in 1000s)	10,020,449	10,084,714	9,883,421	8,901,491
Softwoods: Total Drain (ft³ in 1000s)	12,486,586	14,044,294	14,394,028	14,800,999
Softwoods: Annual Growth (ft³ in 1000s)	12,501,271	14,715,427	15,241,092	15,467,789
Softwood Growth to Drain Ratio	*1.00*	*1.05*	*1.06*	*1.05*
Hardwoods: Annual Mortality (ft³ in 1000s)	1,626,733	2,755,701	3,315,862	4,298,579
Hardwoods: Annual Harvest (ft³ in 1000s)	4,215,500	5,971,328	5,690,561	4,139,708
Hardwoods: Total Drain (ft³ in 1000s)	5,842,233	8,727,029	9,006,423	8,438,287
Hardwoods: Annual Growth (ft³ in 1000s)	9,425,003	10,232,615	11,503,274	9,541,561
Hardwood Growth to Drain Ratio	*1.61*	*1.17*	*1.28*	*1.13*
All Species: Annual Mortality (ft³ in 1000s)	4,092,870	6,715,281	7,826,469	10,198,087
All Species: Annual Harvest (ft³ in 1000s)	14,235,949	16,056,042	15,573,982	13,041,199
All Species: Total Drain (ft³ in 1000s)	18,328,819	22,771,323	23,400,451	23,239,286
All Species: Annual Growth (ft³ in 1000s)	21,926,274	24,948,042	26,744,366	25,009,350
All Species Growth to Drain Ration	*1.20*	*1.10*	*1.14*	*1.08*

TABLE 2.9: HISTORY OF US GROWTH TO DRAIN RATIOS FOR SOFTWOODS AND HARDWOODS

Note also that Growth to Drain ratios can vary dramatically by region and species. In the US South, for example, the Growth to Drain for softwoods is significantly higher than for softwoods in the entire US. **Table 2.10** shows the data that supports that statement, with both naturally regenerated and plantation stands of SYP (i.e., the overwhelming majority of softwoods in the US South), the Growth to Drain ratio is well over 1. This means that each year 80 percent more wood is being added to the standing volume than is be-ing utilized or is dying from natural mortality. As is further discussed in Chapter 3, the combination of a high percentage of privately owned lands and a large amount of excess growth are leading to extensive investment in new sawmilling capacity across the region.

Canadian Growth to Drain

Table 2.11 provides a nearly 30-year history of Growth to Drain for Canadian forests. Note

ACOUSTI-MAT, SOUND CONTROL FOR MASS TIMBER

	NATURAL STANDS	PLANTATIONS	TOTAL
US South: Softwood Annual Growth (ft³ in 1,000,000s)	2,886	5,972	8,858
US South: Softwood Annual Harvest (ft³ in 1,000,000s)	721	3,225	3,946
US South: Softwood Annual Mortality (ft³ in 1,000,000s)	603	323	926
US South: Total Drain (ft³ in 1,000,000s)	1,324	3,548	4,872
Growth to Drain Ratio	2.2	1.7	1.8

TABLE 2.10: US SOUTH GROWTH TO DRAIN RATIO FOR SOFTWOODS IN 2017

that total wood supply is the Annual Allowable Cut (AAC), a calculated value that projects the amount of timber that can be harvested sustainably based on the capacity of the forests to grow new fiber and their natural mortality. As the data shows, in all cases, the actual harvest levels have been lower than the AAC by an average factor of 1.4 for all species, 1.3 for softwoods, and 2.2 for hardwoods. This is a positive finding for the mass timber industry, as it indicates that Canadian forests are not being overharvested and could supply more fiber if increasing market demand warranted it.

2.2.2 ENVIRONMENTAL FOREST MANAGEMENT CERTIFICATION

Many forest landowners manage with multiple objectives in mind and consider sustainability in their forest management planning and decision-making. Environmental forest management certification programs offer landowners a formal process for ensuring that their plans are consistent with sustainability objectives related to fiber production; wildlife habitat; clean water; recreation values; and the wide range of plants, animals, insects, and fungi that make up the web of life in a forest ecosystem.

Concerns about sustainability and the protection of myriad forest values emerged in the United States and Canada during the 1960s, '70s, and '80s. As a result, laws such as the National Environmental Policy Act (NEPA), Endangered Species Act, Clean Water Act, Clean Air Act, National Forest Management Act (NFMA), and others were passed. All these laws help ensure a baseline of sustainability and accountability in forest management, especially on public lands. In the 1990s, however, concerns arose about the sources of wood from private lands and from countries where illegal logging is prevalent or forest management practices are lax.

Those concerns, spurred by buyers of wood products who wanted assurance that their products

YEAR	TOTAL WOOD SUPPLY	TOTAL HARVEST	TOTAL G:D	SOFTWOOD SUPPLY	SOFTWOOD HARVEST	SOFTWOOD G:D	HARDWOOD SUPPLY	HARDWOOD HARVEST	HARDWOOD G:D
1990	8,747	5,523	1.6	6,367	4,986	1.3	2,246	537	4.2
1991	8,687	5,445	1.6	6,371	4,891	1.3	2,182	554	3.9
1992	8,518	5,781	1.5	6,247	5,184	1.2	2,133	597	3.6
1993	8,405	5,989	1.4	6,166	5,315	1.2	2,101	674	3.1
1994	8,408	6,265	1.3	6,145	5,445	1.1	2,129	819	2.6
1995	8,267	6,470	1.3	6,035	5,558	1.1	2,094	908	2.3
1996	8,285	6,282	1.3	6,028	5,343	1.1	2,122	939	2.3
1997	8,373	6,484	1.3	6,078	5,431	1.1	2,161	1,052	2.1
1998	8,295	6,141	1.4	6,028	5,043	1.2	2,175	1,098	2.0
1999	8,454	6,946	1.2	6,169	5,749	1.1	2,186	1,197	1.8
2000	8,281	7,045	1.2	6,099	5,767	1.1	2,140	1,278	1.7
2001	8,369	6,512	1.3	6,215	5,294	1.2	2,147	1,218	1.8
2002	8,415	6,900	1.2	6,254	5,636	1.1	2,161	1,261	1.7
2003	8,472	6,406	1.3	6,289	5,078	1.2	2,179	1,328	1.6
2004	8,730	7,349	1.2	6,540	5,950	1.1	2,182	1,398	1.6
2005	8,641	7,109	1.2	6,431	5,834	1.1	2,207	1,275	1.7
2006	8,733	6,445	1.4	6,547	5,251	1.2	2,179	1,190	1.8
2007	8,881	5,724	1.6	6,696	4,753	1.4	2,186	964	2.3
2008	8,836	4,884	1.8	6,692	4,033	1.7	2,140	844	2.5
2009	8,507	4,089	2.1	6,413	3,330	1.9	2,091	756	2.8
2010	8,362	4,979	1.7	6,314	4,146	1.5	2,041	830	2.5
2011	8,186	5,181	1.6	6,162	4,269	1.4	2,016	911	2.2
2012	8,115	5,269	1.5	6,116	4,400	1.4	1,992	869	2.3
2013	8,023	5,332	1.5	6,053	4,450	1.4	1,967	886	2.2
2014	8,112	5,308	1.5	6,060	4,404	1.4	2,052	901	2.3
2015	8,052	5,488	1.5	5,993	4,524	1.3	2,059	961	2.1
2016	7,875	5,484	1.4	5,791	4,520	1.3	2,084	968	2.2
2017	7,741	5,445	1.4	5,693	4,450	1.3	2,052	996	2.1
2018	7,695	5,516	1.4	5,633	4,531	1.2	2,062	985	2.1

TABLE 2.11: COMPARISON OF ANNUAL ALLOWABLE CUT TO ACTUAL HARVEST IN CANADA (CUBIC FEET IN MILLIONS)

were sourced from well-managed forests, led to the development of environmental forest management certifications. Through the Earth Summit in Rio de Janeiro and the Montreal Process meetings in the early 1990s, forest health and management criteria and indicators were developed. They were to be monitored by independent, third-party verification groups. The intent was to create a market-driven reward for complying with the criteria and indicators judged to represent sound, sustainable forest management. Wood is the only building material that has third-party certification programs in place to demonstrate compliance with sustainability principles.

In the decades since, only about 11 percent of the world's forests have been certified as complying with one of several programs, according to the Global Forest Atlas from the Forest School at the Yale School of the Environment. Despite accounting for only 11 percent of the certified acreage, those certified forests provide an estimated 29 percent of global timber production. More than 92 percent of all certified forestland is in the Northern Hemisphere, with the US and Canada accounting for more than half that total. The acreage of certified land in tropical forests is approximately 2 percent. Thus, even though certification was conceived as a means of stopping deforestation—primarily a tropical forest issue—little forest management has been certified among the world's tropical forests. Note that the species and lumber products produced from tropical forests are not used in the production of mass timber products. Thus, the mass timber industry has little direct impact on tropical forest management and deforestation.

Forest Certification in the US and Canada

Across the US and Canada, more than 480 million acres of forestland, or roughly 20 percent of all North American forests, have been certified under various third-party forest certification programs. The four main certification programs operating in North America are listed here:

- American Tree Farm System: ATFS is managed by the American Forest Foundation and is designed to serve relatively small family forest ownerships. ATFS is endorsed by the Programme for Endorsement of Forest Certification (PEFC), a global umbrella organization that endorses a variety of national forest certification systems. Through ATFS's association with PEFC, ATFS-certified landowners have global certification status. See additional information here: https://www.treefarmsystem.org/.

- Forest Stewardship Council: FSC was initiated in 1993 and is a global forest certification program. As of 2019 (the most recent annual report), nearly 500 million acres have been certified globally. In North America, FSC certificate holders include publicly owned forests, native forest enterprises, family forest trusts, and industrial timberlands. Roughly 160 million acres are FSC-certified in North America. See additional information here: https://www.fsc.org.

- Sustainable Forestry Initiative: SFI was started in 1994 and primarily serves large industrial forest landowners. It is endorsed by PEFC. As of 2019, about 375 million acres of North American forestland have been certified to the SFI standard. See additional information here: https://www.forests.org/.

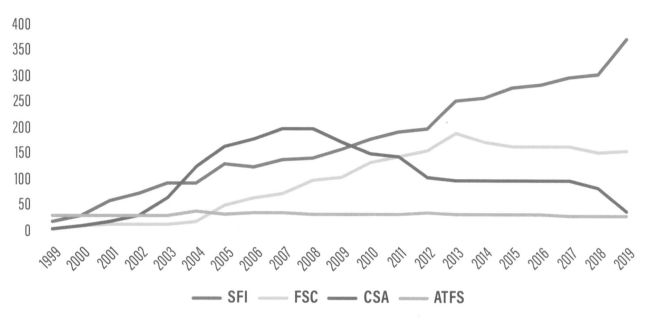

FIGURE 2.4: HISTORY OF ACRES CERTIFIED IN NORTH AMERICA BY FOREST CERTIFICATION PROGRAMS

- Canadian Standards Association: CSA is the Canadian standards system established in 1996. Like SFI and ATFS, CSA is PEFC-endorsed. See more information here: https://www.csasfmforests.ca/.

Figure 2.4 shows the history of the acres certified in North America under each program. Note that data in the figure was interpolated by the author team from a figure included in the *2020 SFI Annual Progress Report*.[3]

Certification of Public Lands in the United States

Most federal land in the United States, including national parks, national forests, Bureau of Land Management lands, and wildlife refuges, is not certified to the standards of any of the above programs. Rather, federal laws guide manage-

ment planning and activities. Large areas of federal land have been permanently set aside from timber harvest. These include wilderness areas, national parks, and inventoried roadless areas. Such reserved areas play an important role in sustainability by providing habitat conditions not always found on forestlands managed for timber production.

Generally, state and municipally owned lands are managed to generate sustained revenue from the harvest of timber and use of other resources. Revenue from management activities often supports school systems and other rural, local government needs. Unlike federal lands, several states and municipal governments have enrolled in one of the above forest management certification programs. Landowners who have not pursued third-party certification are guided by state and municipal laws and/or Best Management Practices (BMPs)

---

3   Accessed at: https://www.forests.org/progressreports/

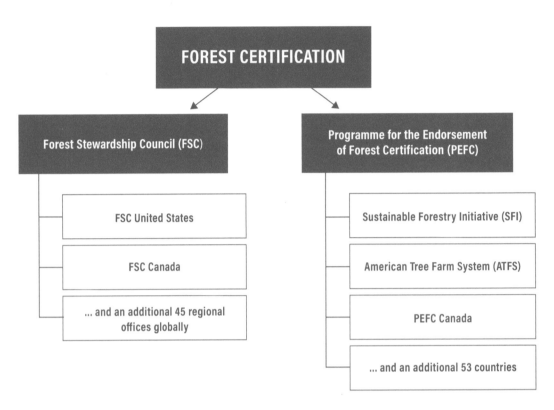

FIGURE 2.5: RELATIONSHIPS AMONG FOREST CERTIFICATION PROGRAMS

that govern or guide forest management within a jurisdiction. The nature and extent of these laws vary considerably across the US. Common to all, though, are principles designed to assure clean water and long-term sustainability. Thus, at a minimum, end users can be assured that forest management in the US overwhelmingly complies with local, regional, and federal forest management laws.

Certification of Public Lands in Canada

Most Canadian forestland is publicly owned. However, a tenure system allows private companies to carry out sustainable forest management on public lands. Under the tenure system, the right to harvest a public resource (timber) is transferred to a private entity. Although details vary from province to province, the basic concept is that a company signs a long-term agreement with the Canadian government. The agreement encompasses a designated forest acreage and dictates certain forest management guidelines (i.e., applicable forestry laws, regulations, and policies) with which the company must comply. In addition to those standards, about 420 million acres of forest in Canada have been certified by third parties, including FSC, SFI, and CSA. Canada also has 59 million acres reserved from harvest in the form of parks and other protective designations. The reserved areas represent about 6 percent of Canada's forests.

Future of Forest Certification

A report[4] released by Dovetail Partners, a non-profit that provides authoritative information about the impacts and trade-offs of environmental decisions, analyzes what the future of forest certification might look like. A key takeaway is that competition among forest certification programs may hinder the ability of forest certification to continue having a meaningful impact on forest management. **Figure 2.5**, adapted from the Dovetail Partners report, shows the divergence among forest certification programs: the FSC program stands alone while the PEFC program acts as an umbrella organization for numerous global programs.

Key drivers cited as threats to the programs are the steady growth within supply chains of private- and public-sector alternative approaches to forest certification, technological innovation, and government policies. The report offers ways to ratchet down competition, including the suggestion that supply chain influencers adopt either a neutral position about material sourced from the different programs or rank their choice to sourcing certified fiber in order of preference. According to Dovetail Partners, ranked choice is an alternative to the "all or nothing" approach that is apparently a common practice among some sectors of end users.

---

4   Dovetail Partners, *Forest Certification Update 2021: The Pace of Change* (January 2021), https://dovetailinc.org/ upload/ tmp/1611160123.pdf.

2.3 FOREST DIVERSITY

Species richness, the number of unique species in an area, is frequently used as a measure of forest sustainability. In the US, there are many different ecological zones, translating into numerous species of trees. During US Forest Service FIA timber cruises in 2017, cruisers identified nearly 1,000 unique species. Most abundant were red maple, loblolly pine, balsam fir, sweetgum, and Douglas fir. However, when considered on the basis of biomass rather than tree count, Douglas fir makes up the largest portion, accounting for about 1 percent of all the aboveground biomass.

Almost all US forests are native species, and most are naturally regenerated, with planted forests accounting for just 10 percent to 15 percent of the total. In the past 25 to 30 years, government agencies and nonprofit groups have warned that some forest types—and the plant and animal species associated with them—are in decline. Coalitions were formed to reverse the declines. In the Eastern US, they include longleaf pine and shortleaf pine restoration efforts. In the West, restoration projects have focused on Western white pine, whitebark pine, quaking aspen, and ponderosa pine. These groups recognize the desirability of restoring native forests and their associated species. Similarly, a parallel program resulted in restocking Central European forests with native hardwoods (earlier replaced by mono culture softwood plantations). The efforts to find a high-end market for these species resulted in commercialization of the high-performance hardwood mass timber products (mainly beech and oak glulam and beech LVL). In Canada, most forests are made up of native species. A little over half of the harvested acreages are replanted, while half rely on natural regeneration. Canada boasts several different forest types.

2.4 FOREST HEALTH

What is a healthy forest? The answer is nebulous, but the primary disturbance agents are clear: insects, disease, and wildfire. How landowners view the impacts of those disturbance agents differs depending on their management goals. If the forest is reserved (i.e., wilderness or a national park) and the purpose is to manage for natural processes, the definition of "healthy" is different than that for land managed by a publicly traded company seeking a return on investment for shareholders. A noncorporate family forestland manager with diverse goals will provide yet another definition. The answers reflect different objectives. What is healthy also varies by forest ecosystem, requiring different management practices.

In reserved forests, insect outbreaks, wildfires, and chronic endemic diseases lead to patterns of high natural mortality followed by natural regeneration. While disastrous from a wood utilization viewpoint, these patterns may be considered healthy from other vantages because they are part of a forest's natural processes. Dead trees, for example, become habitats for birds, plants, mammals, and insects. These natural agents of change can be considered desirable in some forests and undesirable in others—for example, where they destroy valuable timber, damage a municipal watershed, or spoil scenic vistas.

The owner of forests managed for timber production wants to manage tree mortality to reap an economic benefit and provide a renewable product that supports society's need for human habitat in the form of homes, shops, and offices. Some timberlands are managed to blend different objectives. As described earlier, many family forests and public lands are managed for a mixture

FIGURE 2.6: EXAMPLE OF A HIGH-INTENSITY FOREST FIRE

of goals, so some mortality from fire, insects, and diseases may be acceptable and even desirable. Even so, severe die-offs are not desirable. Maintaining a balance is an important part of managing the forest.

2.5 FOREST FIRE RESILIENCE

Forest fires and the smoke they generate once again filled the news in 2021. Wildfire risks are driven by two synergistic factors. As the climate warms and wildfire seasons lengthen, the risk of "megafires" increases. The problem is exacerbated by limited management activity on some ownerships and by 100 years of aggressive wildfire suppression. Forests that once burned frequently now have abnormally large quantities of green and dead trees, and thickets of brush. The fuel buildup is particularly acute in western

North America. High-intensity wildfires are ever more common, with proportionately severe consequences (see **Figure 2.6.**).

Many land managers, scientists, wildfire managers, and increasingly, the public are calling for action to mitigate these risks. Two common treatments are thinning, removal of forest fuels including some trees and underbrush; and controlled burning, intentional burning with a low-intensity fire to reduce ground fuel buildup without damaging the overstory of large trees. Many of the forests in need of treatment are not traditional industrial forestlands. More often, they are public lands and family forests where the public's tolerance for cutting or burning trees is low. Some treatment areas are in municipal watersheds with reservoirs that serve domestic and agricultural water users.

FIGURE 2.7: THE EFFECT OF FOREST MANAGEMENT ON FIRE BEHAVIOR

USDA Forest Service, How Fuel Treatments Saved Homes from the Wallow Fire,
https://www.fs.usda.gov/Internet/FSE_DOCUMENTS/stelprdb5318765.pdf.

Thinning and prescribed burning are both costly because the cost of removing smaller trees is almost always greater than their commercial value. However, when thinning and burning costs are weighed against the immense cost of firefighting and the associated loss of lives, property, and resources, these forest health treatment projects may make sense economically. There are many examples around the country where proactively treating forests saved property, lives, and even communities. For example, **Figure 2.7** shows how forest management affected the Wallow Fire in Arizona. High on the ridge (upper portion of photo), the fire killed the trees as it burned with high intensity through the tree crowns. Lower on the ridge (middle portion of photo), the forest had been thinned prior to the fire, and when the flames reached that area, the fire dropped from the tree crowns and became a much lower-intensity ground fire that allowed the trees to survive and firefighters to prevent the loss of several homes and structures (foreground of photo).

Thinning can be accomplished with mechanical harvesting equipment, by crews sawing trees and piling them for burning, or with planned low- to moderate-intensity burns completed under prescribed conditions. Often, thinning and burning are used in conjunction with each other with greatest efficacy. Some trees in need of removal can be used for forest products, including mass timber. When such markets exist, it becomes considerably more affordable to manage forests for the desired outcomes.

The increased use of mass timber products can expand markets for some small and medium trees that should be thinned to reduce the risks of wildfires, insect outbreaks, and diseases. The use of more wood in commercial buildings helps create new demand, leading to more logging and manufacturing capacity. In addition to the forest health benefits, the increased activity can lead to new jobs in the forest and at manufacturing plants, especially in rural communities with limited opportunities for building viable economies.

2.6   FOREST CARBON

Forests are key to the Earth's natural carbon capture and storage system. During photosynthesis, trees take in carbon dioxide (along with sunlight and water) to create simple carbohydrates, or sugars, that can be used to either nourish their existing cells or create new cells (growth). When used for growth, carbon is stored by creating woody material. When the sugars are consumed for nourishment, the tree releases carbon dioxide as a by-product back into the atmosphere. In the continental US alone, forests store more than 14 billion metric tons of carbon (see **Table 2.12**).

If unaltered by human activity, the complete life cycle of a tree is carbon neutral. However, this cycle can take hundreds of years to complete, depending on local conditions and the species

involved. Some are relatively short-lived (only 80 to 120 years), such as quaking aspen and lodgepole pine. Others can live many centuries, such as ponderosa pine, Douglas fir, Western larch, and others. A forest is often a mix of species with varying life spans and adaptations. Some ecosystems have frequent natural disturbance cycles, only decades apart, and others have cycles lasting centuries. Disturbances come in many forms: fire, insect epidemics, drought, hurricanes, ice storms, windstorms, and more. And many of these interact, creating synergies. A windstorm, for example, can blow down hundreds or thousands of acres of trees to then provide a food base for bark beetles or other insects to breed and expand their populations to then attack live trees. These events can set the stage for high loads of fuel in the forest that can feed a severe wildfire.

The natural, or unmanaged, tree and forest cycles can be thought of as having three phases: carbon capture, carbon storage, and carbon release. The cycle for an individual tree and the overall forest may or may not be synchronous depending on the disturbance regime. In the first phase, a tree grows and uses carbon dioxide absorbed from the atmosphere as its building blocks. In the second phase, the tree is mature and no longer uses as much carbon for growth. Instead, the tree consumes a larger portion of its sugars to maintain its current systems, and so is not as efficient at capturing and storing carbon. In the third phase, the tree releases more carbon than it captures as it declines in vigor and parts of the tree begin to decay. It then dies of old age, disease, insect attack, or fire, eventually releasing most of its remaining carbon back into the atmosphere. A portion will remain in the soil, if undisturbed. In the natural forest, while some trees decline or die, others will regenerate, grow, and replace them, and in the

STATE	NATIONAL FOREST	OTHER FEDERAL	PRIVATE	STATE & LOCAL	TOTAL
AL	21	9	492	19	541
AR	65	18	318	18	419
AZ	70	12	53	7	142
CA	494	79	352	51	976
CO	184	46	53	7	290
CT	0	0	44	18	62
DE	0	0	9	3	12
FL	22	23	191	70	306
GA	32	21	473	23	549
IA	0	3	49	7	59
ID	318	14	48	24	404
IL	8	2	96	11	118
IN	6	5	106	11	129
KS	0	2	40	1	42
KY	28	12	284	8	333
LA	20	13	262	21	316
MA	0	2	69	33	104
MD	0	2	64	22	88
ME	2	4	299	28	333
MI	65	7	254	83	409
MN	38	4	124	83	249
MO	32	7	251	19	309
MS	39	15	408	13	476
MT	270	25	62	14	371
NC	45	25	414	31	516
ND	11	1	7	1	20
NE	0	1	18	1	21
NH	23	2	97	14	136
NJ	0	3	27	26	56
NM	84	12	63	9	169
NV	21	38	2	0	61
NY	1	4	388	157	549
OH	9	2	185	27	223
OK	9	9	120	7	144
OR	539	159	257	52	1,007
PA	19	5	344	144	512
RI	0	0	8	4	12
SC	20	14	255	19	307
SD	0	0	8	1	10
TN	25	23	313	32	394
TX	21	13	421	13	468
UT	71	46	23	11	151
VA	56	19	387	22	484
VT	14	2	105	15	136
WA	349	111	250	131	841
WI	32	5	226	57	319
WV	43	7	327	13	391
WY	81	30	12	3	126
Total	3,189	856	8,659	1,383	14,087

TABLE 2.12: TONS OF CARBON IN FORESTS BY STATE BY OWNERSHIP TYPE (METRIC TONS IN MILLIONS)

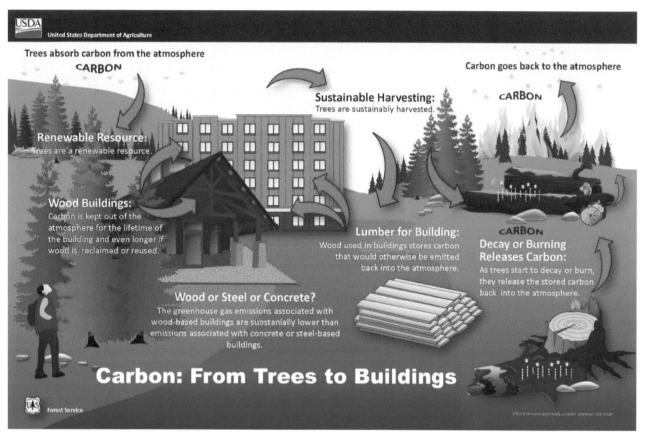

Carbon: From Trees to Buildings

process absorb and sequester more carbon. In a forest with a long disturbance cycle, the dead trees might retain quite a bit of carbon as they slowly decay, or they might release it relatively quickly if the species of wood is more susceptible to rot. If it is a forest with more frequent disturbances, like fire, then the carbon stored in dead wood, litter, and duff is much lower.

As part of actively managing forests, the carbon cycle can be extended. After trees are harvested, they can be manufactured into durable, long-lived products that can continue storing carbon while in service. The harvested forests regenerate with vigorous growth, starting a new cycle. Active forest management often decreases natural mortality and captures usable material before the

carbon release cycle begins. Wood then enters the industrial cycle in the form of products that store carbon in building structures, furniture, packaging, and paper (see **Figure 2.8**).

The carbon sequestration impact of a wood product is contingent on how the forest it comes from is managed. Forest certifications like FSC and SFI help consumers source sustainable materials, but it is often unclear which practices are more effective at achieving the outcomes desired in the market. A lot depends on the kind of forest in question. Ongoing research will help inform the evolution of forest practices in an era of critical carbon sequestration, and it will show how building design teams can incorporate wood into their Life Cycle Analyses (LCA).

Consensus around how to shift practices to a balanced triple bottom line in forestry is desirable at each point in the wood products supply chain, but the path forward is not yet clear. Fortunately, the exponential increase in interest in mass timber products has captured the public imagination in ways not seen since Smokey Bear, pushing a much-needed wave of multidisciplinary conversations around carbon stewardship in forests.

Because forests play such a critical role in absorbing atmospheric carbon, it is important to avoid converting forestlands to other uses. Although it may seem counterintuitive, one way to ensure that forestlands remain forested is to provide an economic return to the landowners. North America and Western Europe have some of the highest per capita wood use in the world, but they also have net positive forest growth. That's because the demand for and value of wood products creates an economic incentive to maintain forests. In developing countries, deforestation is often driven by the desire to produce something more valuable for the landowner, so the land is converted to nonforest uses. Increasing the demand for and value of wood and the forests that produce them reduces the risk of deforestation.

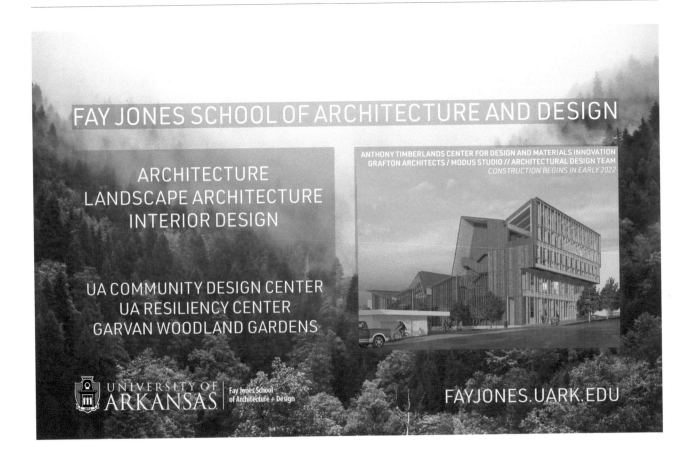

FRERES LUMBER CO. HAS INVESTED OVER $5 MILLION, PLANTING 3 MILLION SEEDLINGS ON FIRE-DAMAGED ACREAGE. IN 40 YEARS, THESE LANDS WILL BE COMMERCIALLY THINNED, AND IT WILL BE ANOTHER 60 TO 80 YEARS BEFORE FRERES SEES RETURNS ON THAT INVESTMENT.

Source: Freres Lumber Co.; Photo Source: Skyline Video Productions

CASE STUDY: FRERES LUMBER

MANUFACTURERS, FORESTS BENEFIT FROM SALVAGED WOOD

LOCATION: LYONS, OREGON

MASS TIMBER MANUFACTURER: FRERES LUMBER CO.

THREE-QUARTERS OF GREENHOUSE GASES emitted by a wildfire are generated by postfire decay, meaning dead trees that rot after a fire, contribute more greenhouse gases than are emitted from the fire itself, according to researcher Dr. Thomas M. Bonnicksen, professor of forestry at Texas A&M University. Removing dead and dying trees and re-planting young trees after a fire helps to rebalance the carbon cycle and fight climate change.

In September 2020, 5 simultaneous megafires burned in Oregon; 1.2 million acres were scorched. Santiam Canyon was hit hard by the Beachie Creek and Lionshead fires, which burned more than 375,000 acres combined. Nine people lost their lives, ecosystems were reduced to ash, forests were incinerated, and countless businesses and proper-ties were destroyed More acres burned in the Cas-cade Range than in any year on record.

Freres Lumber Co. lost 5,760 acres of prime trees in the 2020 infernos. Between cleanup and destroyed timber, Freres's losses will exceed $24 million.

ACRES						
	USFS	**BLM**	**ODF**	**LARGE PRIVATE**	**OTHER PRIVATE**	**TOTALS**
Total Acres	364,300	129,300	25,200	271,400	180,700	970,900
Nonforest and Roads	16,900	6,300	1,100	10,700	16,200	51,200
Forested Acres	347,400	123,000	24,100	260,700	164,500	919,700
Unburned and Low Severity	132,400	43,300	12,600	73,100	63,500	324,900
Medium and High Severity	215,000	79,700	11,500	187,600	101,000	594,800
Premerchantable	53,600	27,000	5,100	137,000	58,600	281,300
Merchantable	161,400	52,700	6,300	50,600	42,400	313,400
Unavailable for Sustainable Harvest	108,700	37,800	400	6,200	3,300	156,400
Available for Sustainable Harvest	52,700	14,900	6,000	44,400	39,000	157,000
Probable Salvage Acres	18,400	11,000	4,800	38,000	33,600	105,800

ACRES BURNED (SORTED BY SEVERITY AND HARVESTABILITY)
Source: Oregon Forest Resources Institute Report, September 2021

WASTE NOT, WANT NOT

If Freres Lumber's own lands are any indication, 95 percent of the burnt volume of timber in these forests is salvageable, but only if it is harvested quickly. Once bugs and decay set in, the fiber loses value. More than half of the salvageable material is on federal land. But little to none of it is being sold because of litigation brought by environmental groups and regulations that limit harvest.

As burnt logs decay, they release methane, which has a comparative impact 25 times greater than carbon dioxide. Dry, dead logs also provide tinder for future fires.

BENEFITS OF SALVAGED WOOD

If burnt logs are quickly harvested and turned into carbon-sequestering wood products, the release of greenhouse gases can be arrested, and carbon can be locked into building materials. About 75 percent of the wood fiber used in Freres products in 2021 came from salvaged wood. All of that fiber is going into Freres's veneer and Mass Ply products for projects that include remodeling the Portland International Airport.

FIGHTING CLIMATE CHANGE

Carbon sequestration is one of the most important strategies in countering climate change. Through the process of photosynthesis, trees are especially good at taking in and storing carbon. Researchers have found that active forest management can reduce the intensity of wildfires and improve the health of our forests and ecosystems. Salvaging trees for wood products and quickly replanting is one facet of active forest management that likely will help keep forests healthy while fighting increasing levels of carbon dioxide. in our atmosphere. ◯

CHAPTER 3: RAW MATERIALS

IMPACTS OF THE MASS TIMBER EFFECT

- It is estimated that each square foot of building constructed with mass timber consumes, on average, 0.9 cubic feet of mass timber raw material (panels and beams).

- Each cubic foot of finished mass timber (panels) is estimated to require 22.5 board feet (nominal tally) of dimensional lumber to produce.

- US and Canadian softwood lumber production in 2020 was about 60 billion board feet. There is ongoing significant investment in softwood lumber sawmilling capacity in the US South.

Mass timber is a somewhat unusual market for sawmillers because the lumber must be dried to a lower moisture content than lumber used in other applications. Because kiln drying is often the bottleneck in a given sawmill's output capacity, the sawmiller's ability and willingness to do "extra" drying is an important factor in mass timber's raw material supply chain. It's a fact: the manufacture of mass timber requires raw material that, for most products, is dimension lumber made from various softwood species. Those interested in mass timber will find it helpful to understand the key features of these raw materials. Accordingly, this chapter includes a technical analysis of the specifications for use in mass timber; a look at the production capacity among raw material manufacturers (e.g., sawmills); and an estimation of the demand for raw materials that mass timber's development could create for suppliers.

FIGURE 3.1: ILLUSTRATION OF A MASS TIMBER PANEL'S MAJOR (PARALLEL OR LONGITUDINAL) AND MINOR (PERPENDICULAR OR TRANSVERSE) STRENGTH DIRECTIONS

3.1 RAW MATERIAL SPECIFICATIONS

The following sections summarize the specifications for sawn lumber and Structural Composite Lumber[1] (SCL) used in mass timber products. More detailed information is available in the design standard reference specific to each product type.

3.1.1 CROSS-LAMINATED TIMBER (CLT)

Before launching into a technical discussion about how lumber can be used in mass timber, we'll first discuss the terminology. Every CLT panel has major and minor strength axes. The major axis is the direction with the greatest number of layers of wood grain in a parallel orientation. For example, **Figure 3.1** shows a 3-layer panel. The grain of the wood in the 2 outer layers is *parallel*, and thus the longest axis of the panel is the major strength direction. Sometimes the parallel axis is also called the *longitudinal* axis. The wood grain in the middle layer is oriented

---

1   Structural Composite Lumber (SCL) is a family of engineered wood products that includes Laminated Veneer Lumber (LVL), Parallel Strand Lumber (PSL), Laminated Strand Lumber (LSL), and Oriented Strand Lumber (OSL). These products are created by combining wood veneers, wood strands, or wood flakes with moisture-resistant adhesives to form blocks of material known as billets. The billets are then sawn into sizes roughly analogous to sawn lumber.

perpendicular to the adjacent layers. Because there is only 1 perpendicular (or *transverse*) layer, it is the panel's minor strength direction. The following technical sections reference these italicized terms.

The Engineered Wood Association (APA) developed a standard that addresses the manufacturing, qualification, and quality assurance requirements of CLT panels. It's called *ANSI/APA PRG 320 – 2019: Standard for Performance-Rated Cross-Laminated Timber.* The most recent edition was approved by the American National Standards Institute (ANSI) on January 6, 2020.

Section 6, Subsection 6.1 of *ANSI/APA PRG 320* is the portion of the standard that specifies the characteristics of the sawn lumber and SCL that are approved for use in CLT panels. The following list summarizes key aspects; see the PRG 320 report for full details.

Species

Specific to the North American mass timber market, lumber from any softwood species[2] or species combination (e.g., hem-fir, fir-larch, spruce-pine-fir) recognized by the American Lumber Standards Committee (ALSC) under PS 20 or by the Canadian Lumber Standards Accreditation Board (CLSAB) under CSA-0141 with a minimum published specific gravity of 0.35 is permitted. Any given layer (lamination) in a CLT panel shall be made from lumber of the same thickness, type, grade, and species or species combination. Ad-

jacent layers in a CLT panel can be made from differing thicknesses, types, grades, and species or species combinations. If SCL is made from any species with a specific gravity greater than 0.35 and meets the standards of ASTM[3] D5456, it is permitted. Finally, note that strict enforcement of species and grade restrictions in panels imported from overseas manufacturers as integral project parts may not be practical or even desirable.

Lumber Grade

The distinction between major and minor strength axes is important because differing lumber grades are required, depending on whether they are in a longitudinal or transverse layer. Lumber is graded in one of two ways: (1) visually—where strength/grade is estimated from a visual inspection, or (2) machine stress rated (MSR)—where pieces of lumber are measured for resistance to bending and assigned a strength rating. In a CLT panel's longitudinal layers, the lumber grade must be visual grade No. 2 (or better), or MSR grade 1200f-1.2E. Perpendicular layers must be at least visual grade No. 3 or the equivalent. Any proprietary lumber grades meeting or exceeding the mechanical properties of the approved CLT lumber grades can be used if they meet an approved agency's qualifications.

Thickness

The minimum thickness of any lumber layer in a CLT panel is ⅝ inch (16 millimeters) at the time of gluing. Maximum thickness is 2 inches (51 millime-

---

2   The higher a species' specific gravity, the more dense the wood; and generally, the more dense the wood, the greater its strength properties. Douglas fir, larch, Western hemlock, Southern Yellow Pine (SYP), lodgepole pine, Norway pine, various spruce species, and various true firs are common North American softwoods that have good strength properties.

3   ASTM International, formerly known as the American Society for Testing and Materials, is an international standards organization that develops and publishes voluntary consensus technical standards for a wide range of materials, products, systems, and services.

LONGITUDINAL LAYERS				TRANSVERSE LAYERS			
Nominal Size (inches)	Actual Thickness (inches)	Actual Width (inches)	Ratio (Actual Width to Actual Thickness)	Nominal Size (inches)	Actual Thickness (inches)	Actual Width (inches)	Ratio (Actual Width to Actual Thickness)
1x2	0.75	1.5	2	1x2	0.75	1.5	2
1x3	0.75	2.5	3.33	1x3	0.75	2.5	3.33
1x4	0.75	3.5	4.67	1x4	0.75	3.5	4.67
1x6	0.75	5.5	7.33	1x6	0.75	5.5	7.33
2x2	1.5	1.5	1	2x2	1.5	1.5	1
2x3	1.5	2.5	1.67	2x3	1.5	2.5	1.67
2x4	1.5	3.5	2.33	2x4	1.5	3.5	2.33
2x6	1.5	5.5	3.67	2x6	1.5	5.5	3.67
2x8	1.5	7.25	4.83	2x8	1.5	7.25	4.83
2x10	1.5	9.25	6.17	2x10	1.5	9.25	6.17
2x12	1.5	11.25	7.5	2x12	1.5	11.25	7.5

TABLE 3.1: ALLOWABLE AND UNALLOWABLE THICKNESS-TO-WIDTH RATIOS FOR LUMBER USED IN CLT PANELS
Any cell in red font is a lumber size with a thickness-to-width ratio that renders that size unacceptable for use in CLT panels.

ters) at the time of gluing. Thickness must be consistent across each individual layer. Thickness consistency is defined at the time of bonding as plus or minus 0.008 inch (0.2 millimeters) across the width of the layer, and plus or minus 0.012 inch (0.3 millimeters) across the length of the layer. Per PRG 320, any bow or cup present in lumber "should be small enough to be flattened out by pressure in bonding." Many overseas national or regional lumber markets offer much broader selections of thicknesses. The overseas CLT manufacturers take advantage of that variety by offering panel lay-ups more efficiently adjusted to project requirements. Some CLT manufacturers in Central Europe utilize laminations as thin as 0.4 inches (10 millimeters). These lay-ups would not meet the PRG 320 minimum lamination thickness requirement.

Width

For longitudinal layers, the net lamination width for each board shall not be less than 1.75 times the net lamination thickness. For transverse layers, the net width of a board shall not be less than 3.5 times the net thickness of the board. **Table 3.1** illustrates the thickness-to-width ratios for the longitudinal and transverse layers of common lumber sizes. Note that it is common practice for CLT manufacturers to plane about $^1/_{16}$ inch off all four sides of a piece of lumber prior to panel lay-up. Thus, the thickness-to-width ratios of a board's final dimensions may differ slightly from those shown in the table. Notably, 2-by-4, a common size in North America, cannot be used in transverse layers. Exceptions to these thickness-to-width ratios are allowed if the pieces in a layer are both face- and edge-glued. Laminations made from SCL are permitted to be full CLT width.

Moisture Content

The moisture level of lumber used in CLT panels must be 12 percent, plus or minus 3 percent, when the panel is manufactured. Because lumber shrinks or swells as it loses or gains moisture, the lumber's moisture content is a key focus area for mass timber manufacturers. It is also an important part of the manufacturing process because the majority of lumber is sold after it has been kiln-dried. The grading rules require that lumber be dried to a minimum of 19 percent moisture content. Given these circumstances, sawmills may be reluctant to reduce kiln capacity by running batches of "mass timber lumber" for longer-than-normal drying cycles when demand for lumber is strong. This issue is further discussed in Section 3.4 and from the perspective of the mass timber panel manufacturer in Chapter 4.

Surfacing

Any sawn lumber used in a CLT panel must be planed or sanded—at least on any surfaces to be bonded—and the planed or sanded surface must not have any imperfections that might adversely affect the bonding process (e.g., raised grain, torn grain, skip, burns, glazing, or dust). ANSI and the APA, noting the intricacies of bonding the layers in a CLT panel, state that the bonding surfaces on some species need to be planed within 48 hours of the bonding process. Planing or sanding of face-bonding surfaces of SCL used to make CLT panels is not required unless it's needed to meet thickness tolerances.

3.1.2   NAIL-LAMINATED TIMBER (NLT)

The International Building Code (IBC) recognizes NLT as a structural material and provides guidance on structural design and fire safety. No product-spe-

cific ANSI standard has been developed, but design guides are available for both the US and Canada, and they can be downloaded for free at www.thinkwood.com. NLT is commonly manufactured at the building site by nailing pieces of lumber together after they have been arranged so that their wide faces are touching. Almost any properly graded softwood dimension lumber can be used to make NLT. However, considerations such as cost, availability, species, structural performance (grade), and aesthetics come into play when selecting material. Most NLT panels manufactured to date utilize No. 2 grade dimension lumber in 2-by-4, 2-by-6, and 2-by-8 sizes. The lumber's moisture content must be below 19 percent before fabrication.

3.1.3   DOWEL-LAMINATED TIMBER (DLT)

The structural design of each lamination in a DLT panel is covered by both the IBC and the National Building Code of Canada (NBC). The *International Code Council Evaluation Service Report ESR-4069*, published in November 2020, provides guidance on the use of DLT, given the material's structural and fire-resistance properties. The report evaluates DLT's compliance with the 2018, 2015, 2012, and 2009 IBC and the 2018, 2015, 2012, and 2009 *International Residential Code*. Additionally, StructureCraft, a North American mass timber manufacturer of DLT, has developed a design guide.

Species and Grades

DLT panels are made from spruce, pine, fir (SPF); Douglas fir; and hem-fir species or species groupings. Panels made from other species are available on request. The structural grades include Select Structural; No. 2 and Better; 2400f-2.0E MSR for Douglas fir; and 2100f-1.8E MSR or 1950f-1.7E MSR for SPF.

Moisture

Lumber must be kiln-dried to 19 percent or less moisture content at the time of manufacture. Note that the hardwood wooden dowels used to join the DLT laminations are at a much lower moisture content at the time of manufacture. When the drier dowels are exposed to the wetter softwood laminations, they gain moisture, swell, thereby forming a tight connection between laminations.

Thicknesses and Widths

From a global perspective, in Europe Massiv-Holz-Mauer (MHM) and dowel-bonded CLT favor thinner (nominal 16 millimeter to 25 millimeter, equivalent to 5/8 inch to 1 inch) and wider (200 millimeter and more, equivalent to 8 inches). Both technologies can, however, accommodate rough (undressed) lumber. MHM uses rough-sawn boards rather than nominal 2-by stock. The surface is not considered for visual quality. That means there should be greater potential for utilizing lower-quality lumber than that required for adhesive-bonded CLT. The process favors wider laminations (200 millimeter and more, equivalent to 8 inches). Laminations are grooved on one side along the grain to increase thermal insulation. The final thickness of grooved laminations is about 16.5 millimeters (5/8 inch). Dowel-bonded CLT also uses rough sawn lumber in core layers, but it needs dressed lumber for the face layers, which often are meant to be visible in structures. Also, bonding with dowels requires wide-face lumber (likely more than 200 millimeters, or about 8 inches) to form two rows of successful dowel bonds in each surface layer. This likely limits the prospect of utilizing small logs.

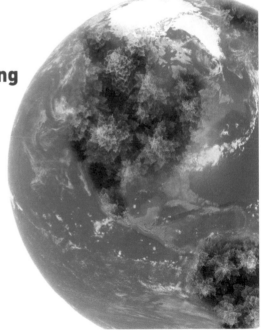

GRADES	PREMIUM	SELECT	STANDARD	INDUSTRIAL
COMMON APPLICATION	Residential; Hotels; Feature Walls	Residential; Libraries; Schooles; Museums; Offices	Offices	Non-Visual; High Ceilings
SPECIES	SPF; Douglas fir; (other species available)			
COATINGS	Upon request, a penetrating clear sealer and tinted top coatings can be shop-applied to exposed side of panel. Our team focuses on working with designers to determine the best coating system for durability and ease of maintenance.			
WANE*	Width < or = 1/4"; Length < or = 2'; No bark	Width < or = 3/8"; W/O bark length < or = 5'; W bark length < or = 2'; Max 1 in every 5 boards	Width < or = 1/2"; W/O bark length < or = 10'; W bark length < or = 7'; Max 1 in every 4 boards	Permitted
KNOTS	No open knots; tight knot permitted	Open Smoth < or = 3/4" diameter; Open Jagged < or = 1/2" diameter; Tight knot permitted	Permitted	Permitted
BLUE STAIN	Max 1 every 10 boards; Up to 10% surface area; No dark/black coloring	Max 1 every 7 boards; Up to 15% surface area; No dark/black coloring	Max 1 every 5 boards; Up to 20% surface area; No dark/black coloring	Permitted
CHECKS IN STRAND EDGE	Non-permitted	Width < or = 1/16"; Length < or = 12"	Width < or = 1/16"; Length < or = 24"	Permitted
CHARACTERISTICS DISTRIBUTION	Distributed	Distributed	Some distribution	No re-distribution required
PANEL SURFACE	Deviation on board-to-board elevation < or = 1/8"		Deviation on board-to-board elevation < or = 1/4"	Deviation permitted
UNNATURAL BELMISHES	Except for Type 4 (Industrial), the underside of the DLT panel shall be free of "unnatural" characteristics, eg. black marks, scuffs, damage, glue, etc. Such blemishes shall be sanded/required as needed.			
CHARACTER OF WOOD	All wood, as a natural material, will exhibit characteristics such as knots/holes, wane, grain, checks, coloration, etc. The intent of the above appearance grading is to provide a degree of predictability/limitation to these characteristics. However, some variations in the visual appearance will be apparent.			

TABLE 3.2: LUMBER CHARACTERISTICS ALLOWED WITHIN STRUCTURECRAFT DLT PANEL GRADES

Wane is the presence of bark or lack of wood fiber along the edge of a piece of lumber.

Lumber Size

DLT panels come in thicknesses ranging from 4 inches to 12.17 inches. Lumber widths are available from 2 inches to 6 inches (nominal).

Appearance

StructureCraft has developed four grades of DLT panels: Premium, Select, Standard, and Industri-al. **Table 3.2** specifies the lumber characteristics of each of StructureCraft's grades.

3.1.4 GLULAM

ANSI A190.1-2017, Standard for Wood Products—Structural Glued Laminated Timber, and ANSI 117-2020, Standard Specification for Structural Glued Laminated Timber of Softwood

SPECIES GROUP	SPECIES INCLUDED IN GROUP
ALASKA CEDAR	Alaska Cedar
DOUGLAS FIR-LARCH	Douglas Fir, Western Larch
EASTERN SPRUCE	Black Spruce, Red Spruce, & White Spruce
HEM-FIR	California Red Fir, Grand Fir, Noble Fir, Pacific Silver Fir, Western Hemlock, & White Fir)
PORT ORFORD CEDAR	Port Orford Cedar
SOUTHERN YELLOW PINE	Loblolly Pine, Longleaf Pine, Shortleaf Pine, Slash Pine
SPRUCE-PINE-FIR	Alpine Fir, Balsam Fir, Black Spruce, Engelmann Spruce, Jack Pine, Lodgepole Pine, Norway Pine, Norway Spruce, Red Spruce, Sitka Spruce, White Spruce
SOFTWOOD SPECIES	Alpine Fir, Balsam Fir, Black Spruce, Douglas Fir, Douglas Fir South, Engelmann Spruce, Idaho White Pine, Jack Pine, Lodgepole Pine, Mountain Hemlock, Norway Pine, Norway Spruce, Ponderosa Pine, Sitka Spruce, Sugar Pine, Red Spruce, Western Larch, Western Red Cedar, White Spruce

TABLE 3.3: SOFTWOOD SPECIES (OR SPECIES GROUPINGS) COMMONLY USED IN GLULAM TIMBERS

Species, are the two documents published by APA that describe the specifications of lumber to be used in glulam timbers.

Key specifications include the following:

Species

The ANSI A190.1-2017 standard states that any softwood or hardwood species is approved for use in structural glued laminated timber if stress indices and knot distributions are established as described in ASTM D3737. The ANSI 117-2020 standard is more specific about allowable species or species groupings, as shown in **Table 3.3.**

Moisture Content

The moisture content of lumber used in glulam timbers shall not exceed 16 percent at the time of bonding.

Wane

Wane is a defect in a piece of lumber characterized by bark or insufficient wood at a corner or along an edge. For dry-service conditions (i.e., when lumber in use is not regularly exposed to moisture), wane up to one-sixth the width at each edge of interior laminations is permitted in certain grade combinations. When that is the case, the basic shear design value shall be reduced by one-third. When wane is limited to one side of a member, the basic shear design value is reduced by one-sixth. Other instances of wane are allowed, but the circumstances are complicated. See ANSI 117-2020 for details.

Grade

Lumber used in glulam timbers is graded visually or mechanically, and it is identified by grade before bonding. Rules approved by the Board of Review of the ALSC or written laminate grading rules apply to visually graded lumber. Rules approved by the Board of Review of the ALSC or special rules that conform with the ANSI A190.1 standard apply to mechanically graded lumber. An accredited inspection agency oversees the qualification of proof-graded lumber, subjecting it to full-size tension tests as set forth in the American Institute of Timber Construction (AITC test 406). A number of more specific grading rules apply, depending on the position of the piece in the glulam timber, its species, whether the lumber is ripped prior to bonding, and other factors. See A190.1-2017 for details.

Bonding

All bonding surfaces—including face, edge, and end joints—are smooth and, except for minor local variations, free of raised grain, torn grain, skips, burns, glazing, or other deviations that might interfere with the contact of sound wood fibers.

Thickness

Laminations are not to exceed 2 inches in net thickness, unless a gap-filling adhesive is used for face and edge bonds.

Dimensional Tolerances

At the time of bonding, variations in thickness across the width of a lamination shall not exceed plus or minus 0.008 inches (2 millimeters). Variations in thickness along the length of an individual piece of lumber or along a lamination shall not exceed plus or minus 0.012 inches (3 millimeters).

3.1.5 POST AND BEAM

Traditionally, post and beam construction utilizes timbers of at least 6 inches in nominal width and thickness. Less guidance is available for the specification of lumber (timbers) for this category of mass timber than for others. Nevertheless, the *Code of Standard Practice for Timber Frame Structures* developed by the Timber Framers Guild (TFG) (www.tfguild.org) in 2018 does provide some. A few specifications follow:

Grade

Grades are Select Structural, No. 1, or No. 2. All structural timbers are to be graded by an approved lumber grading agency certified grader or an individual who has completed a timber grading training course. The lumber grader is to provide a grade stamp or a certificate of grade for each piece of timber. Knots and other natural timber features shall not be construed as defects unless their magnitude exceeds the limits prescribed in the applicable lumber grading rules. Checks are a natural feature resulting from ordinary timber drying and seasoning. Checks that develop after the timber frame has been raised are not construed as defects.

Species

Acceptable species include Douglas fir, Eastern white pine, red oak, white oak, Southern Yellow Pine (SYP), and Alaska yellow cedar.

Moisture

Timbers are to be dried to a maximum moisture content of 19 percent.

Size

Timbers 8 inches by 12 inches and smaller are to be Free of Heart Center (FOHC). Timbers larger than 8 inches by 12 inches are to be boxed heart. All timber sizes are nominal (actual dimensions are typically ½ inch smaller than the nominal size in both thickness and width). Pith is the center of a tree that extends along its long axis. The wood around the pith typically is not as strong as wood nearer to the bark. Therefore, a quality factor for smaller timbers is that the pith (heart center) should not be included in the timbers, or that the pith be "boxed" in the center of the timber on larger timbers.

Surfacing

Timbers may be surfaced four sides (S4S), rough sawn, or hewn. "Surfaced four sides" refers to timbers that are planed smooth an all four sides; rough sawn is timber that has been sawn but not planed; and hewn timbers have been shaped using an axe or other similar tool.

3.1.6   HEAVY TIMBER DECKING

Specifications for heavy timber decking are less prescriptive than other mass timber products. Some guidance is provided in *Heavy Timber Construction,* published by the American Wood Council (AWC). Key excerpts include the following:

Grading

The lumber used in heavy timber framing and decking must be graded in accordance with the rules customarily used for the species. These are generally regional grading agencies, including the Northeastern Lumber Manufacturers Association (NELMA), California Redwood Inspection Service (RIS), Southern Pine Inspection Bureau (SPIB), West Coast Lumber Inspection Bureau (WCLIB), Western Wood Products Association (WWPA), and the Canadian National Lumber Grades Authority (NLGA).

Sizing

The decking used in heavy timber floor decks is to be of sawn or glued laminated plank, splined, or tongue-and-groove planks not less than 3 inches (nominal) in thickness or of planks not less than 4 inches (nominal) in width when set on edge. Splining and tongue-and-groove refer to protrusions and indentations on the sides of lumber pieces so that adjacent lumber pieces can be interlocked. For roof applications, the timbers are to be sawn or glulam, splined, or tongue-and-groove plank not less than 2 inches (nominal) in thickness or of planks not less than 3 inches (nominal) in width when set on edge.

3.1.7   VENEER

Veneer-based mass timber products are ANSI/APA PRG 320-certified and include mass plywood panels up to 11 feet 10 inches wide, 12 inches thick, and 48 feet long. Freres Lumber Co. can also manufacture beams and columns made from veneer up to 12 inches wide, 72 inches deep, and 48 feet long that are ANSI/APA PRG 320-certified.[4]

The veneers are first formed into Laminated Veneer Lumber (LVL) billets (an LVL plank of standardized size) that are then made into Mass Ply-

---

4   The equipment Freres uses to make the columns and beams can handle widths up to 24 inches. Freres is working to achieve APA certification for the larger widths. Additionally, columns and beams can be produced up to 60 feet in length, but current production of longer beams is limited by the length of Freres's press.

wood Panels (MPP). Because the veneers are first formed into LVL billets, certification of MPP falls under the classification of SCL, which includes LVL and is covered under ASTM D5456.

More specifically, the manufacturer uses wood veneers to make LVL billets. These billets are 1.6E, 1.55E, or 1.0E Douglas fir LVL as recognized by APA in the product report PR-L324; the billets also are in accordance with custom lay-ups of ANSI/APA PRG 320 that employ product qualification and mathematical models that use principles of engineering mechanics. The LVL billets can range in thickness from 1 inch to 24 inches and in width from 1.5 inches to 72 inches. Depending on the billets' dimensions and MPP design needs, the billets are parallel laminated, bonded with qualified structural adhesives, and pressed to form a solid panel (i.e., MPP).

Freres uses Douglas fir veneers classified by moisture content and a grade (G1, G2, G3) that's dependent on their strength, as measured by Ultrasonic Propagation Time (UPT) testing, which correlates the time it takes for sound to pass through wood veneers with key strength determinants, such as specific gravity and modulus of elasticity (ratio of stress to strain). Freres is also a manufacturer of veneer and plywood and thus controls the raw material supply for its MPP manufacturing operations from standing timber through the finished product.

3.2 NORTH AMERICAN LUMBER SUPPLY

As the number and size of mass timber construction projects grows, the capacity of sawmills to supply lumber is of considerable interest. Thus,

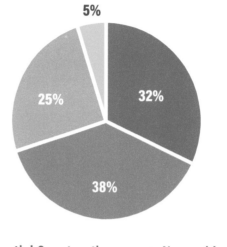

Residential Construction   **Nonresidential**
Repair & Remodeling   **Industrial**

FIGURE 3.2: LUMBER CONSUMPTION BY END-USE MARKET SEGMENT (2016 TO 2020)
Source: Forest Economic Advisors

this section focuses on softwood lumber production and use in North America.

3.2.1 END USES FOR SOFTWOOD LUMBER

Historically, softwood lumber has been used in four key end-use market segments: residential construction, repair and remodeling, nonresidential construction, and industrial/other. **Figure 3.2** shows the average portion of softwood lumber consumed by each end-use market segment in the US from 2016 to 2020. As the data shows, for the last 5 years, on average, 38 percent of all softwood lumber consumed was for repair and remodeling, followed by nearly 32 percent for residential construction. Thus, historical demand has been tied to either new home construction, or the repair and remodeling of existing homes. The industrial/other end-use segment is lumber typically used for applications such as packaging, pallets, and furniture, which utilize the lower grades of lumber. The advent of mass timber and

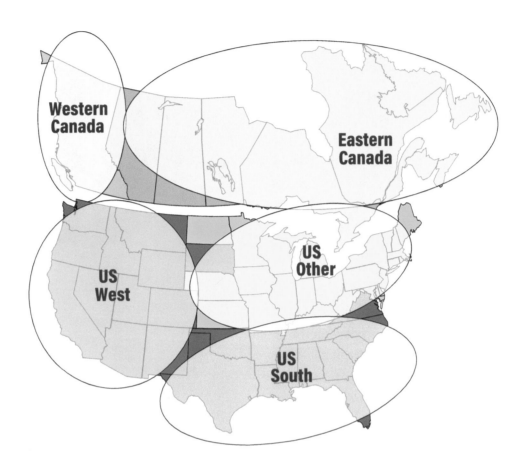

FIGURE 3.3: NORTH AMERICAN SOFTWOOD LUMBER PRODUCING REGIONS

the new demand it places on softwood lumber is the focus of the remainder of this chapter.

3.2.2 WHERE SOFTWOOD LUMBER IS PRODUCED IN NORTH AMERICA

Softwood lumber in North America is produced in five geographical regions: US West, US South, US Other, Western Canada, and Eastern Canada, as shown in **Figure 3.3.** Note that in the US South, 4 species of pine (loblolly, longleaf, short-leaf, and slash) are commonly manufactured into lumber and sold as a species grouping designated as SYP. In Eastern Canada and Western Canada, the predominant lumber grouping is SPF, but the

makeup of species within the SPF lumber grouping differs. In the US West, the predominant lumber species or species groupings are Douglas fir, Douglas fir-larch, and hem-fir.

The volume of softwood lumber produced in each North American region from 2000 to 2021 as reported by the WWPA (2021 is a forecast based on the first six months of the year) is shown in **Figure 3.4.** Note that there is some difference in lumber production and consumption (i.e., some lumber produced hadn't been sold when data was collected). To simplify the discussion, we treat production and consumption as being equal because the volume in

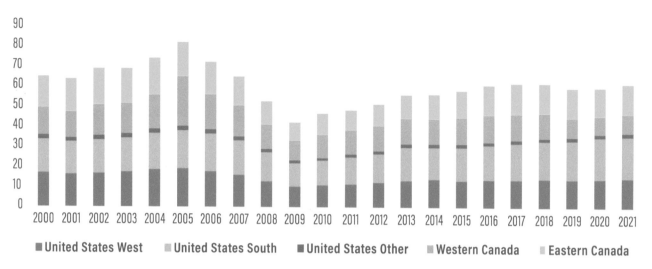

FIGURE 3.4: HISTORICAL UNITED STATES AND CANADIAN SOFTWOOD LUMBER PRODUCTION BY REGION (BOARD FEET IN BILLIONS)

Source: Western Wood Products Association

inventory is typically a small portion of total annual production.

Several things to note about the data in the figure:

- North American softwood lumber production peaked in 2005 at more than 82 billion board feet. At that time Western Canada was the top producing region with nearly 25 billion board feet. At the time, the US West and US South were nearly equal in production, with about 19 billion board feet each.

- Lumber production across North America decreased dramatically during the Great Recession, with totals in 2009 dropping to about 50 percent of the 2005 peak.

- After the low in 2009, North American lumber production increased steadily through 2018 but has since been relatively flat.

- Several regional issues of note:

Western Canada

One of the most dramatic changes is that Western Canada went from producing about 30 percent of North America's lumber in 2005 to producing only an estimated 15 percent in 2021. That change is mainly driven by reductions in the annual allowable cut of timber in the interior region of British Columbia. In that region, a massive mountain pine beetle epidemic affected nearly 45 million acres and killed nearly 60 percent of the standing pines. The outbreak started in the 1990s, and during the 2000s, timber harvests were significantly increased to salvage the dead timber.

The salvage efforts are complete, but current and future harvests have been significantly reduced to allow the forest to grow to a standing inventory that will once again allow for higher levels. Rebuilding is a long process, meaning reduced timber harvest rates will remain in place for the foreseeable future. The sawmill industry built up during the salvage period, and the existing capacity became too large for the available log supply. As a result, many sawmills have permanently closed.

More recently, in late 2021, British Columbia announced plans to defer timber harvests on nearly 200,000 hectares of forests identified as old-growth. If fully implemented, the plans will further reduce lumber production in the region. However, the full impacts remain to be seen as a variety of stakeholders have provided feedback about the announced old-growth deferral harvest plans.

US South

Perhaps equally dramatic are the changes occurring in the US South. Prior to the Great Recession, the US South and the US West produced roughly equal amounts of lumber each year. Since the Great Recession, however, the US South has bounced back. It has recently produced 19 billion to 20 billion board feet of lumber per year—levels that exceed even the peak year of 2005.

A key driver is the 30-year timber harvest rotation brought about by improved forest practices (e.g., genetic improvement of seedlings, and extensive planting and thinning operations). The result is higher volume yields per acre. Thus, during the significant yearslong drop in lumber production during the Great Recession, a massive amount of sawtimber inventory built up "on the stump."

In addition, about 85 percent of the timber belongs to private landowners, meaning sawtimber harvest levels are largely dictated by economic drivers, rather than regulatory drivers. These conditions have spurred massive capital investment in new sawmilling capacity through upgrades to existing mills and greenfield (i.e., new mill at a new site) sawmill development. Nearly 5 billion board feet of capacity has or will come online by 2023. The Beck Group estimates the capital investment that's been made at

$2.5 billion. The nearly 5 billion board feet of new/upgraded capacity and associated capital investment are a "first wave." This is because, despite the increased capacity, there are still regions with excess sawtimber supply. Thus, companies are planning further investments in sawmilling capacity.

US West

Lumber production in the US West increased to almost 15 billion board feet in 2021 after having been essentially flat at 14 billion board feet annually since 2014. The increase in production was spurred by lumber prices that reached unprecedented heights, driven by a prolonged period of strong demand combined with the industry's limited ability to rapidly increase production. Thus, while lumber production did increase in 2021, it was still not enough to meet strong demand driven by new home construction and strong repair and remodeling activity. Many homeowners sought to update their homes to better accommodate both living and working in them.

Another long-standing issue in the US West is that log supplies are constrained by who owns the timberland. Privately held timberland accounts for about 70 percent of the total harvest. Industrial timberland owners manage their timberlands intensively and generally harvest near the maximum allowable sustainable rates. Thus, harvests on industrial lands cannot increase without beginning to deplete the supply of standing timber. Small private timberland owners hold significant acreage and contribute a considerable portion of the annual harvest. This segment, however, is made up of many thousands of individuals and families. As a group, these landowners typically do not act in sync because individuals have a variety of management objectives and timber pro-

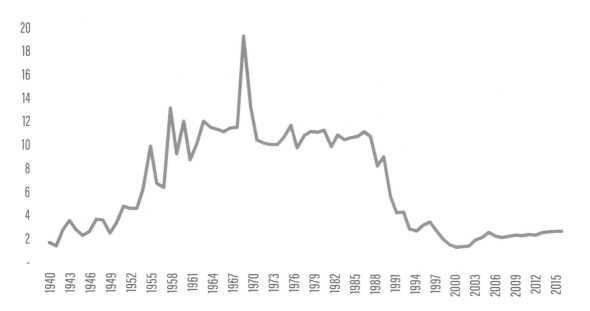

FIGURE 3.5: HISTORY OF US FOREST SERVICE TIMBER SALES FISCAL YEAR 1940 TO FISCAL YEAR 2017 (ANNUAL VOLUME BOARD FEET [LOG SCALE] IN BILLIONS)

Source: https://www.fs.fed.us/forestmanagement/products/cut-sold/index.shtml

duction is not always a top priority. They could supply additional logs, but because they do not act collectively, their supply is constrained.

The balance is under public ownership: the US Forest Service, the Bureau of Land Management (BLM), and miscellaneous states, counties, and municipalities). About 70 percent of all timber-land acres are publicly owned, a high percentage relative to public ownership in the US South. For about the last 30 years, forest management policies on federally owned public lands have constrained log supplies across the US West and limited lumber production.

For nearly 4 decades starting in the mid-1950s, the US Forest Service sold 10 billion to 12 billion board feet of logs each year. The passage of the Endangered Species Act (ESA) required changes to federal policies, resulting in the listing of the northern spotted owl, various salmon species,

and the marbled murrelet. That led to a dramatic decline in the annual volume of timber sold since 1988, as shown in **Figure 3.5.** US Forest Inventory and Analysis (FIA) data suggests that despite the massive tree mortality from the many wildfires in recent years, federal lands are growing 3 times more wood fiber than is being removed by harvesting and natural mortality. That suggests timber harvests could be increased significantly without endangering the sustainability of the resource. In the meantime, increased lumber production is largely held in check by limited log supply.

Eastern Canada

Like other North American regions, except the US South, lumber production in Eastern Canada has been stagnant for the past 5 years, hovering between 14.5 billion and 15 billion board feet per year. Unlike Western Canada, where timber supply is constrained by the lingering effects of the

REGION	% DIMENSION (2" NOMINAL)	ESTIMATED 2021 PRODUCTION OF DIMENSION (BBF)	% SMALL TIMBERS (3"- 5")	ESTIMATED 2021 PRODUCTION OF SMALL TIMBERS (BBF)	% LARGE TIMBERS (6"+	ESTIMATED 2021 PRODUCTION OF LARGE TIMBERS (BBF)	% OTHER	ESTIMATED 2021 PRODUCTION OF ALL OTHER SIZES	TOTAL 2021 PRODUCTION (BBF)
US West	55%	8.1	5%	0.7	5%	0.7	35%	5.2	14.8
US South	80%	16.9	10%	2.1	5%	1.1	5%	1.1	21.1
US Other	20%	0.3	n/a	n/a	n/a	n/a	80%	1.4	1.7
Western CA	75%	7.2	n/a	n/a	n/a	n/a	25%	2.4	9.6
Eastern CA	50%	7.3	n/a	n/a	n/a	n/a	50%	7.3	14.6
North America Total		39.8		2.8		1.8		17.3	61.7

TABLE 3.4: ESTIMATED NORTH AMERICAN SOFTWOOD LUMBER THICKNESS MIX IN 2021 (BOARD FEET IN BILLIONS)

mountain pine beetle, standing timber is readily available. However, parts of Eastern Canada are a long distance from markets, and the average small tree size in those parts increases sawmill manufacturing costs because productivity is constrained by small average piece size. In general, larger mills enjoy economies of scale, allowing for lower manufacturing costs. Historically, Eastern Canada has concentrated on pulp and paper mills that produce newsprint. Those pulp mills were largely supplied with residue from sawmills. As demand for newsprint dwindled, producing lumber from small logs has become more difficult economically, constraining milling capacity.

The smaller tree size also means that lumber tends to be narrower and shorter. To produce a reasonable annual lumber volume, the mills must operate their lines at very high throughput rates (i.e., a large number of logs through processing equipment per unit of time), and because of limited ability to increase feed speeds and still main-

tain good lumber quality, they likely have little ability to increase those rates. However, in late 2020, Quebec announced plans to nearly double the amount of wood harvested by 2080. In the near term, timber harvest is expected to increase from 29 million cubic meters per year to 33 million cubic meters by 2025 and reach nearly 53 million cubic meter/year by 2080. These changes, if implemented, will translate into higher levels of timber production in Quebec. See https://mffp. gouv.qc.ca/wp-content/uploads/FS_overview_ QTPS.pdf for more details.

Global

Forests in Central Europe are still suffering through a widespread spruce bark beetle outbreak. The trees killed are being salvaged, but that has created a glut of logs and dampened lumber prices. As a result, European mass timber manufacturers are enjoying low raw material costs relative to their North American peers,

REGION	% ABOVE #2	ESTIMATED BBF ABOVE #2	% OF #2	ESTIMATED BBF OF #2	% OF #3	ESTIMATED BBF OF #3	% BELOW #3 AND OTHER	ESTIMATED BBF OF BELOW #3 & OTHER	TOTAL PRODUCTION OF DIMENSION (BBF)
US West	35%	2.8	55%	4.5	5%	0.4	5%	0.4	8.1
US South	40%	6.8	40%	6.8	10%	1.7	10%	1.7	16.9
US Other	10%	0.0	55%	0.2	20%	0.1	15%	0.0	0.3
US Total		9.6		11.4		2.2		2.1	25.3

TABLE 3.5: ESTIMATED US SOFTWOOD DIMENSION LUMBER GRADE MIX IN 2021 (BOARD FEET IN BILLIONS)

making European producers cost competitive. There is also anecdotal evidence that Central European and Baltic mass timber manufacturers are increasingly balancing their raw material needs with lumber produced in Russia. As discussed in more detail in Chapter 4, global lumber market conditions during 2021 have also meant that European manufacturers have supplied much of the US mass timber market.

3.2.3 2020 NORTH AMERICAN SOFTWOOD LUMBER PRODUCTION DETAILS

As described in Section 3.1, mass timber product standards specify the use of certain lumber sizes and grades. Thus, the grades and sizes of lumber produced are also important. **Table 3.4** shows lumber production by thickness, based on the WWPA's estimated North American softwood lumber production volumes for 2021. The percentages of production by size values are estimates from sawmill industry benchmarking data collected by The Beck Group. Of the estimated

61.7 billion board feet of lumber produced in North America in 2021, about 65 percent is nominal 2-inch-thick dimension lumber (i.e., boards nominally 2 inches thick and 8 feet to 20 or more feet long). Of the remainder, only small portions are made into thicker timbers, and another 25 percent or so are in other, miscellaneous sizes. Note that most of the volume in the "other" category is stud-grade lumber. It is the same thickness as dimension lumber, but it is produced only in 4-inch and 6-inch widths, and mainly in lengths of less than 12 feet. Most stud-grade lumber is used as vertical structural components in wall systems for homes. The balance of the "other" category includes industrial and common boards (i.e., nonstructural lumber), and miscellaneous products.

Similarly, it is useful to understand the grade mix, as shown in **Table 3.5**. Using the WWPA's 2021 production estimates and The Beck Group's sawmill benchmarking data, the table shows that nearly 85 percent (21 billion board feet out of 25.3 billion board feet) of dimension lumber pro-

REGION	% 2-BY-4	ESTIMATED 2-BY-4 PRODUCTION (BBF)	% 2-BY-6	ESTIMATED 2-BY-6 PRODUCTION (BBF)	% 2-BY-8	ESTIMATED 2-BY-8 PRODUCTION (BBF)	% 2-BY-10	ESTIMATED 2-BY-10 PRODUCTION (BBF)	% 2-BY-12	ESTIMATED 2-BY-12 PRODUCTION (BBF)	TOTAL 2020 DIMENSION PRODUCTION (BBF)
US West	40%	3.2	30%	2.4	10%	0.8	10%	0.8	10%	0.8	8.1
US South	25%	4.2	30%	5.1	20%	3.4	15%	2.5	10%	1.7	16.9
US Other	40%	0.1	30%	0.1	10%	0.03	10%	0.03	10%	0.03	0.3
US Total		7.6		7.6		4.2		3.4		2.5	25.3

TABLE 3.6: ESTIMATED US SOFTWOOD DIMENSION WIDTH MIX IN 2021 (BOARD FEET IN BILLIONS)

duction in the US is No. 2 grade or better. Data for Canada is not included because the information was not readily available, but the grade yields are likely similar.

Finally, **Table 3.6** displays the estimated width mix. As the data shows, about 30 percent of all dimension lumber is estimated to be 4 inches wide, followed by another 30 percent that is about 6 inches wide. A significantly higher percentage of 2-by-4s is produced in the US West than in the US South. Lumber width is a significant consideration for mass timber manufacturers, as prices vary among widths, and productivity improves when wider pieces of lumber are used.

3.2.4 SOFTWOOD LUMBER PRICING

The purchase of raw material is the single largest cost associated with the manufacture of mass timber products, accounting for more than 50 percent of a plant's total operating cost. Lumber pricing, therefore, is a key focus area for manufacturers. In the US over the past 10 years, demand for lumber in the residential construction and repair and remodeling market segments has ranged from a low of 20.8 billion board feet per year to a high of 34.5 billion board feet per year. The associated swings in supply create considerable volatility in lumber prices, a phenomenon that is unusual in the rest of the world because in many countries lumber is less commonly used to construct homes.

Price volatility was in full swing in 2021. For about the past 25 years, the price of dimension lumber in North America has averaged roughly $350 per 1,000 board feet (MBF). The low point occurred in 2009 in the depths of the Great Recession when dimension lumber was selling for around $200 per MBF. The high point was in mid-2018 when prices approached $600 per MBF. In 2021, however, driven by a COVID-induced increase in demand in the home repair and remodeling sector and by a constrained ability to produce lumber because of COVID-related labor shortages and other supply constraints related to shipping, prices skyrocketed to all-time highs. By June 2021, dimension lumber prices in North America averaged more than $1,500 per MBF, more than four times the long-term average price.

3.2.5 ENVIRONMENTAL CERTIFICATION OF SOFTWOOD LUMBER

In Chapter 2 of this report, we explained how forested land is certified when managed under certain protocols that have been judged to represent sustainable forest management. Such forest management programs also offer chain-of-custody certification to participants in the supply chain. Chain-of-custody is the process of certifying that, as products move from the forest to the end user, material originating from certified forests is identified or kept separate from noncertified material. Chain-of-custody certification generally involves detailed logistics and materials-handling protocols, inventory management, batch processing, filings, and third-party audits.

Forest management and chain-of-custody certification fulfills the end users' desire for assurances that the products they are using are from well-managed forests. This is especially true for developers seeking to certify a building under Leadership in Energy and Environmental Design (LEED) and similar programs. In addition, large tech companies like Google and Facebook that have expressed interest in mass timber are keenly interested in using environmentally certified raw materials. But it isn't yet clear which environmental certification programs these large and influential mass timber users will prefer.

Forest landowners and wood product manufacturers who follow the forest management and chain-of-custody guidelines can market their products as being environmentally certified. It is difficult to track the volume of lumber (and veneer/plywood) sold annually in North America that is environmentally certified because a high percentage of these forest products could be environmentally certified under one or more of the programs. But frequently they are not marketed in that manner, and thus there is no well-documented record of their sales volumes.

One of the main reasons these sales are not well-tracked is that, for most consumers, this attribute is relatively unimportant. Considerations such as price, quality, species, and grade are much more important. In addition, landowners and manufacturers must expend considerable effort and money to acquire and maintain these certifications. Given the limited market demand and the expense, many landowners and manufacturers decide not to certify their products, even though they could be. Others elect to certify their material on a case-by-case basis as dictated by customer expectations. A small number of producers choose to certify as much of their product as possible, regardless of the level of demand from customers.

For producers of mass timber products, this means that market demand for environmentally certified materials—aside from mass timber products—is relatively low. Therefore, finding environmentally certified material may be an challenge, but likely it is not a roadblock. In interviews with The Beck Group, the general feeling of mass timber producers is that only a small portion of their demand is for certified mass timber products, and when those orders need to be filled, they can usually oblige. It may cost more, however, to acquire cer-

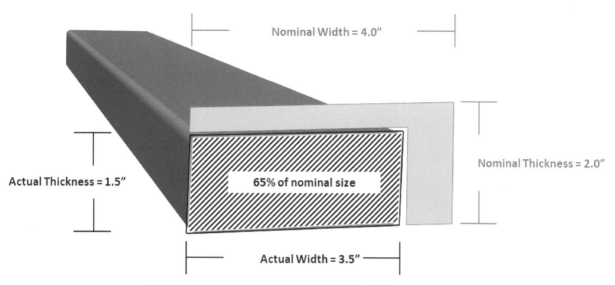

Nominal Width = 4.0"

Nominal Thickness = 2.0"

Actual Thickness = 1.5"

65% of nominal size

Actual Width = 3.5"

FIGURE 3.6: COMPARISON OF NOMINAL AND ACTUAL DIMENSIONS FOR BOARD FOOT LUMBER TALLY

Source: The Beck Group

tified lumber. As previously noted, a big wild card is whether one of the large tech companies will announce plans for a large mass timber project (or projects) and give preference to raw materials from a given environmental certification program. Such an event would likely trigger a rise in the price of environmentally certified raw materials until the supply chain is able to adjust to the increased demand.

3.3 THE MASS TIMBER INDUSTRY'S ESTIMATED DEMAND FOR RAW MATERIALS IN 2021

In this section, we provide an estimate of the lumber demand arising from mass timber products. First, however, it is important for readers to understand a quirk of the North American lumber industry—the difference between actual and nominal lumber sizes.

3.3.1 NOMINAL VERSUS ACTUAL LUMBER SIZES

As described in Chapter 1, an estimated 22.5 board feet (nominal tally) is needed to produce 1 cubic foot of finished mass timber panel. Some readers may be thinking that 22.5 board feet per cubic foot seems like too much lumber input per cubic foot of finished panel. Such thoughts likely stem from the knowledge that a board foot is defined as 1 inch thick by 12 inches wide by 12 inches long. Thus, it may seem that one cubic foot of mass timber should be equal to 12 board feet.

This is not the case for several reasons. First and most importantly, softwood lumber in North America is bought and sold on a nominal board foot basis. For example, a common lumber size is 2 inches thick by 4 inches wide. Those dimensions, however, are nominal, meaning in name only. The actual dimensions are 1.5 inches thick by 3.5 inches wide. As shown in **Figure 3.6**, this means that about 35 percent of the

Lumber Size (Thick x Width)	ACTUAL			NOMINAL			Actual Fiber % (Actual/ Nominal)	Air Space %
	Actual Thickness (Inches)	Actual Width (Inches)	Cross-Sectional Area (Inches²)	Nominal Thickness (Inches)	Nominal Width (Inches)	Cross-Sectional Area (Inches²)		
2-by-4	1.5	3.5	5.25	2	4	8	65.60%	34.40%
2-by-6	1.5	5.5	8.25	2	6	12	68.80%	31.30%
2-by-8	1.5	7.25	10.88	2	8	16	68.00%	32.00%
2-by-10	1.5	9.25	13.88	2	10	20	69.40%	30.60%
2-by-12	1.5	11.25	16.88	2	12	24	70.30%	29.70%

TABLE 3.7: COMPARISON OF THE PERCENTAGE OF ACTUAL FIBER TO AIRSPACE AMONG LUMBER SIZES FOR NOMINALLY TALLIED LUMBER

area in a 2-inch by 4-inch space is air. Because so much of a tally is airspace, more than 12 board feet of lumber will be needed to produce a cubic foot of mass timber panel. Additionally, about 8 percent to 10 percent of a board's thickness is planed away before it is glued up. Planing activates the wood surface for the adhesive to bond it. Also, during finger-jointing, a portion of the incoming lumber becomes waste as defects are cut out with a chop saw. Finally, a portion of a mass timber panel is lost to trim around the perimeter and cutouts for windows, doors, and other openings.

The percentage of airspace decreases as lumber width increases, as shown in **Table 3.7**. Nevertheless, a significant portion of the board foot tally for every piece of lumber is airspace.

3.3.2 ESTIMATED LUMBER CONSUMPTION BY MASS TIMBER IN NORTH AMERICA

Definitive data about lumber consumption among mass timber producers is not readily available. However, an analysis completed by the Softwood Lumber Board[5] suggests that in the near term, lumber demand associated with mass timber can reach about 1 billion board feet per year and could grow to nearly 5 billion board feet per year by 2035. This topic is analyzed in further detail in Chapter 4.

3.4 SUPPLYING THE MASS TIMBER MARKET: SAWMILLER PERSPECTIVES

Conceptually, sawmillers are always interested in developing new markets for the lumber they

5   Softwood Lumber Board, *Mass Timber Outlook*, https://softwoodlumberboard.org/wp-content/uploads/2021/03/SLB-Mass-Timber-Outlook-2021-Final-Condensed.pdf.

produce. However, dimension lumber is a largely commoditized product in North America. As such, prices are volatile as various supply and demand factors ebb and flow. Regardless, manufacturers face the constant discipline of producing at a low cost. Thus, many sawmillers tend to operate their mills in a manner that emphasizes high productivity and minimizes distractions that slow production without adding significant value.

For the mass timber market, the area where the sawmiller's mindset has had the largest impact is lumber drying. As previously stated, the specification for lumber used in mass timber is 12 percent, but the grading rules for kiln-dried dimension lumber only require drying to 19 percent moisture. Thus, lumber destined for mass timber manufacturing must receive extra drying at the sawmill, or the mass timber manufacturer must have a means of further drying the lumber at their facility.

From the perspective of the sawmiller, lumber drying is often the "bottleneck" in the manufacturing process. In other words, the output of the entire operation (i.e., sawmill and planer mill) is limited by the capacity of the dry kilns. Taking extra time to dry lumber to a lower moisture content, therefore, is a decision that must be carefully considered because it takes extra time in the kilns, and the yield of lumber of the appropriate sizes and grades must be considered. In other words, not all the lumber in any given batch sent through a kiln will meet mass timber's grade and size specifications. That lumber is known as downfall (i.e., material that cannot be used for mass timber manufacturing). Thus, some percentage of the lumber that receives the additional time and expense of extra drying cannot be sold to mass

timber manufacturers and must be used in other, less stringent applications.

The strategies for dealing with this issue are evolving. In one approach, sawmilling companies have contracted with mass timber manufacturers to provide lumber that meets their moisture content specifications. Such contracts likely include a significant above-market premium on the price to account for the extra drying time and downfall. Data isn't available on the premium, but The Beck Group estimates it is likely in the range of $50 to $100 per MBF. But during lumber market cycles when prices are high, as it was during parts of 2020 and 2021, numerous sawmilling companies are unwilling to slow their process, regardless of any premium.

In another emerging strategy, mass timber manufacturers are purchasing "ordinary" kiln-dried lumber that may have air-dried during shipment and storage to an acceptable range for mass timber manufacturing. Such an approach requires that the manufacturer have an inline moisture meter in their manufacturing process, allowing the sorting of boards that can be used from those that have too much moisture. The "wet" boards are then set aside. They can be diverted to an on-site controlled drying process or an off-site custom drying service, or they can be set aside for more air-drying, an uncontrolled process.

Each has advantages and disadvantages as described below:

- **On-site, controlled kiln-drying:** Some mass timber manufacturers have invested in their own kilns, so the moisture content issue can be addressed in a controlled manner and with their own equipment. The advantage of this approach is that it allows for the best control over product quality. The disadvantage is that expenses increase. They include the up-front capital expense of kilns; the ongoing operating costs of the kilns (both labor and energy); and the potential yield loss from any material that degrades during the kiln-drying process (e.g., case-hardening, bowing, cupping, warping) to the point it can no longer be used for mass timber. Plus, an experienced kiln operator is needed. Some mass timber manufacturers report that they are experimenting with dehumidification kilns, which operate at much lower temperatures, meaning less likelihood of degradation. The drawback is that drying takes longer at the lower temperatures. Early results indicate that because relatively little moisture needs to be removed from lumber already kiln-dried to 19 percent moisture, the slow drying issue is mitigated.

- **Off-site, custom kiln-drying:** Some mass timber manufacturers have used the services of off-site, custom kiln-dryers. The advantage of this, like on-site, is that it allows for the drying of lumber under controlled conditions. The disadvantages are the costs for handling the wet lumber, paying the kiln-drying service, and transporting the lumber to and from the custom kiln-drying site. In addition, the availability of custom kiln-drying services differs by region.

- **Uncontrolled air-drying:** Lumber will lose or gain moisture depending on ambient air conditions. Thus, lumber simply left to further air dry may lose enough moisture to be used in mass timber panels. There is no energy cost, but this uncontrolled process depends on weather conditions. Thus, it may work only during certain times of the year and in certain regions where ambient conditions generally allow for drying. Also, for best results, the lumber should be placed on stickers (i.e., spacers between layers of lumber that allow air flow and, in turn, drying). The labor and time associated with the handling increase the expense.

In yet another approach, some mass timber manufacturers are part of vertically integrated companies that have sawmilling, kiln-drying, and mass timber manufacturing capacity. This trend applies particularly to the mass timber panel industry and is not specific to North America, but observed in many other mass timber producing regions. The advantage from a mass timber manufacturer perspective is that the material is controlled—often from the tree in the forest (when timber is either owned or purchased standing)—through the manufacture of a mass timber panel. Assuming kiln-drying capacity is not a bottleneck at such operations, the issue of moisture content is less problematic. During the lumber market conditions experienced in 2020, vertically integrated operations are better able to "hold the line" on raw material costs in the production of mass timber panels, but they must recognize that doing so comes at the opportunity cost of selling lumber in an extraordinarily hot market at prices never before seen in the softwood lumber industry.

3.5 CARBON CONSIDERATIONS

The September 2017 issue of *Forest Products Journal* included an article[6] that analyzed the

---

6   Michael Milota and Maureen E. Puettmann, "Life-Cycle Assessment for the Cradle-to-Gate Production of Softwood Lumber in the Pacific Northwest and Southeast Regions," *Forest Products Journal* 67, no. 5/6.

carbon impact associated with the production of softwood dimension lumber in the Pacific Northwest and Southeastern US. Key conclusions from the study were that the global warming impact indicator is that 129 pounds of carbon dioxide equivalent equivalent was released for each cubic meter of lumber produced in the Pacific Northwest, and 179 pounds of carbon dioxide equivalent for each cubic meter of lumber produced in the Southeastern US. An additional key finding was that in the Pacific Northwest, nearly 1,900 pounds of carbon dioxide equivalent equivalent are stored per cubic meter of lumber produced, and nearly 2,100 pounds of carbon dioxide equivalent equivalent are stored per cubic meter of lumber in the US South. Thus, there is a net carbon benefit of nearly one ton of carbon dioxide equivalent equivalent associated with wood use for the duration of the product's useful life.

These findings are a stark contrast to other common building materials (e.g., steel and concrete) that do not store any carbon dioxide equivalent equivalent during their useful life and that require considerable energy and associated carbon emissions be expended in their manufacture. The study also notes that in lumber production, well over 90 percent of the global warming impact arises from the process of manufacturing (e.g., sawing, planing, kiln-drying, and packaging). Only a very small percentage of the impact arises from the energy expended in log processing and transport (i.e., forest operations).

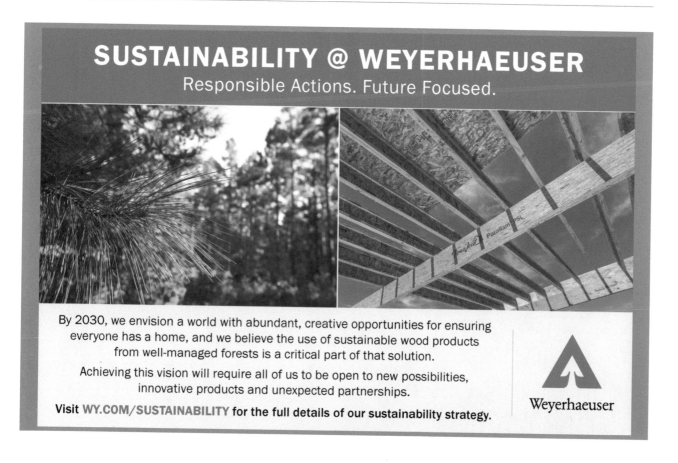

Brains *and* Brawn

Freres' Mass Ply is a veneer-based engineered wood product used to create monolithic panels, beams, and columns. These components are prefabricated to exact dimensions, simplifying transportation and assembly, while making a variety of concepts possible for modern multi-level structures. They are strong, fire-resistant, and lighter per volume than many traditional building materials. As a sustainable and renewable wood product, they're eco-friendly, using trees as small as 8" in diameter to produce panels as large as 12' x 48' x 12". They actually sequester carbon over their lifespan, and compared to concrete and steel, require a fraction of the energy to produce.

》 **Renewable and sustainable**

》 **Stress-tested and rated**

》 **Prefabricated to exact dimensions**

》 **Sequestors carbon over lifespan**

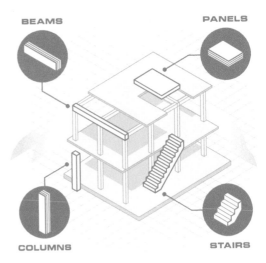

CHAPTER 4: MASS TIMBER PANEL MANUFACTURING

IMPACTS OF THE MASS TIMBER EFFECT

- The estimated practical annual capacity of the North American mass timber industry is 1.04 million cubic meters, which is virtually no change in estimated practical capacity over the prior year. This is because the idling of one facility was offset by the addition of others.

- COVID-related slowdowns in building projects and a COVID-related run-up in North American lumber prices created difficult market conditions for North American mass timber manufacturers in 2020 and 2021.

- Mass timber manufacturers continue to refine their services and means of bringing product to market, including adding staff with timber engineering/design expertise, establishing partnerships, developing design guides, and creating supporting businesses that better link the mass timber manufacturing and building construction communities.

This chapter focuses on mass timber panel manufacturing. Included is a review of the manufacturing processes for key mass timber panels; a list of North American manufacturers and their production capacities, products, and services; and a discussion of strategic and technical mass timber manufacturing issues.

4.1 MASS TIMBER PANEL TYPES

There are two basic types of mass timber panels: those for use in buildings and those for use as industrial matting. Each is described in more detail in the following sections.

4.1.1 BUILDING PANELS

Manufacturers have developed two common building panel grades based on appearance rather than strength: architectural grade, for use when a panel surface will be exposed to the building's occupants; and industrial grade, for use when a panel surface will be covered or will not show. Either grade can be PRG 320[1] certified, if needed. Each manufacturer offers an array of finishes; in most cases, the finishes can be customized.

Architectural-grade panels are designed to ensure that the lumber is of the proper grade and species for visual exposure. The panels may require special sanding, epoxy finishes, stains, or coatings; and the filling of holes, gaps, or knotholes. In addition, lumber grain orientation may be varied. The panel's face layer typically does not include physical defects; to accomplish that objective, an appearance-grade layer of lumber (hardwood or softwood) may be laminated on. Each manufacturer offers its own set of architectural-grade finishes.

Industrial-grade panels are likely to have the same strength characteristics as comparable architectural-grade panels, but they may not meet the same aesthetic standards because the

---

1   "ANSI/APA PRG 320: Standard for Performance Rated Cross Laminated Timber," https://www.apawood.org/ ansi-apa-prg-320.

surface of the panel is usually covered when it's installed. Visual defects on the face layer of industrial-grade panels may include unfilled voids on the edge of laminations, loose knotholes on face layers, or wane (lumber pieces that are not fully square-edged on all four corners). Industrial-grade panels are typically less expensive than architectural-grade panels because the costs of materials, labor, and machining are lower.

In addition, the panel's application plays a significant role in its grade type. A floor may have architectural grade on the bottom side to serve as an exposed ceiling in the room below but industrial grade on the top side because a floor covering will be installed over the top side. Similarly, many exterior walls will be covered with a siding; therefore, only one face of the panel may be architectural grade. Mass timber panels used in roofs and elevator shafts are typically industrial grade.

4.1.2 MATTING

Matting panels are intended not for use in buildings, but to protect the environment and in other industrial uses. These mats typically are placed on the ground to form temporary roads to prevent environmental degradation caused by the heavy machinery used in mining, drilling, pipelines, utility right-of-way maintenance, and remote construction. Traditionally, mats are made of lower-value hardwood timbers nailed or bolted together. Cross-Laminated Timber (CLT) mats are becoming more common because they offer superior value. They are lighter and have longer useful life spans. CLT mats also may include built-in hardware—making them easier to lift and place using a forklift, excavator, or crane and thereby reducing setup time. Matting panels and their uses are described in more detail in Section 4.6.

4.2 MASS TIMBER PANEL MANUFACTURING PROCESS DESCRIPTIONS

Each of the following subsections describes the basic manufacturing steps for key mass timber panels.

4.2.1 CROSS-LAMINATED TIMBER

CLT is produced in an industrial-scale, dedicated manufacturing facility. Although CLT is an innovative product, the major steps in its manufacturing process use well-established technologies borrowed from other segments of the wood products industry although some like lay-up and pressing, are performed on assemblies of an unprecedented scale. Though many major variations are being practiced, the basic manufacturing process typically includes:

Raw Material Receiving

Lumber is received into inventory at the mass timber manufacturing facility.

Raw Material Preparation

Lumber is sorted by grade, width, and species, and moisture content is checked to ensure that the variation among pieces is not too great. Pieces that are too wet are separated for additional drying. Excessive defects (e.g., knots, wane) on the lumber pieces designated for the manufacturing process are removed using a crosscut/chop saw.

Finger Jointing

Once free of excessive defects, the pieces of lumber are glued together end to end, using a machine that cuts finger joints into the lumber ends

FIGURE 4.1: ILLUSTRATION OF FINGER-JOINTED LUMBER
Source: The Beck Group

and applies an adhesive to the joint to securely bond the pieces. **Figure 4.1**.

Cutting to Length

The finger-jointing process creates a "continuous" piece of lumber that can be cut to any length that is called for by the dimensions of the mass timber panel. These range from 4-foot to 12-foot lengths for the panel's transverse axis and 30-foot to 60-foot lengths for its longitudinal axis.

Surfacing

This process, also known as "planing," removes a small amount of material (typically about 1/16 of an inch) from all 4 sides of the piece of lumber. This gives all pieces the same dimensions and activates their surfaces to assure good absorption and bonding of the adhesive used to glue the panel layers together. In particular, the thickness has to be within 0.2 mm tolerance of adjacent pieces to avoid bridging over thinner laminations in cross-laminated lay-ups. Such bridging will result

in an inadequate pressure and poor bond integrity in that location.

Panel Lay-Up

The finger-jointed, surfaced, and cut-to-length lumber pieces are assembled into a panel one layer at a time. In a 3-layer panel, for example, all the long pieces that make up the longitudinal axis are assembled. Next, a glue spreader travels over them, applying a layer of glue to the wide surfaces (note that for some panels, glue is applied to all four sides). Then the short pieces are assembled into the layer making up the panel's transverse axis. Another layer of glue is applied. And finally, the long pieces making up the second major axis layer are assembled. As global perspective, some manufacturers apply adhesive on the narrow edges of laminations as well and include pressure in 2 or 3 directions to produce effective edge-bonding. A substantial volume of panels fabricated by a few major manufacturers in Europe begin the process with edge-bonding the layers before they are used to build lay-ups ready for face-bonding.

Pressing

After the adhesive has been applied and the lumber has been formed into a lay-up, it is pressed while the adhesive cures. Several variations on the adhesive and pressing technology affect the press time and the amount of energy consumed. The majority of the processes use glue that does not require heat to cure, which makes the press times longer. Some processes use glue that needs heat to activate, which reduces press time. Heated presses use radio frequency waves to penetrate the panel and cure the glue. Since adding radio wave energy complicates the process, these types of presses typically cure the lay-up piecewise, one segment at a time.

Final Manufacturing

When the mass timber panels come out of the press, their edges are typically irregular and overrun by adhesive that has bled out between the layers. In addition, the "raw" panels are slightly oversized. All of this means that the panel is cut to its final dimensions in a secondary process. Typically, the final manufacturing is accomplished with a Computer Numerical Control (CNC) machine. Most CLT plants may use a sander to surface the visible face of the panel if architectural finish is required.

Packaging and Shipment

The final step involves the placing of "pick points," metal hardware that allows cranes at the construction site to pick up a panel and place it in a building. For shipping, panels are assembled into a sequence so that when they are delivered to the construction site, they can be moved directly into place rather than having to be unloaded and stored.

The equipment needed to complete the preceding tasks includes the following:

- **Moisture meter:** Tests the moisture content of each piece of lumber, ensuring that any lumber not meeting the target range (12 percent +/- 3 percent) is rejected.

- **Optical grade scanner:** Photoelectric sensors, also known as "photo eyes," that identify any lumber with unacceptable defects (rot, splits, wane).

- **Defect trim saw:** Cuts out the short, lineal sections of lumber identified for removal by grade scanning.

- **Finger jointer:** Cuts finger joints in the ends of each piece of lumber, applies glue to each joint, and presses the pieces together, making one continuous piece.

- **Crosscut saw:** Cuts the finger-jointed lumber to lengths appropriate for the final size of the CLT panel. The only limits on the length of a CLT panel are size of the press and the highway/truck restrictions on the delivery of panels from the manufacturer to the building site.

- **Planer or molder line:** Removes a thin layer of wood from the surface of the lumber to ensure all pieces are of uniform thickness and to "activate" it so it can react to the glue. This step must be completed less than 48 hours before applying the glue.

- **Panel lay-up station:** Arranges pieces of lumber into layers in accordance with the CLT panel design. Glue is applied to each layer at this step. The level of automation varies greatly between operations.

- **Pressing**

  > **Hydraulic press:** Uses hydraulic pressure on face and sides to hold a panel in place as glue cures. Press time varies based on glue formulation and panel lay-up time.

  > **Vacuum press:** Uses a clamshell and silicone blanket to encapsulate a panel and then sucks out the air to tighten gaps between boards.

- **CNC finishing center:** Uses computer-controlled saws and router heads to precisely trim the edges of each panel and cut openings as needed for doors, windows, utility channels, etc.

- **Sanding machine:** Puts a smooth finish on the surface of the panel.

4.2.2 DOWEL-LAMINATED TIMBER PANEL MANUFACTURING

DLT is produced in a dedicated manufacturing facility. As with CLT, incoming lumber is checked for grade and product consistency, and defective sections are removed. The lumber is then finger jointed, cut to the desired lengths, and molded/planed to the desired thicknesses. The cut-to-length boards are assembled into a panel, holes are drilled along the edges, and dowels are pressed into the holes. The entire panel is surfaced to ensure the dowels are not protruding. In the final steps, panels are finished on a CNC machine, packaged, and shipped. Unlike CLT, all the lumber in a DLT panel is oriented in the same direction. This orientation means that the DLT panels do not have the same shear strength properties as those derived from cross-lamination.

4.2.3 NAIL-LAMINATED TIMBER PANEL MANUFACTURING

Unlike CLT and DLT, NLT can be manufactured either at a building site or at an industrial-scale production facility. The layout of an NLT panel is similar to a DLT panel, with all the lumber oriented in the same direction. The lumber is stacked on its side with randomly staggered joints, or it can be finger jointed to create continuous layers over 20 feet long. The boards are then nailed together in various lay-up configurations to create panels.

Industrial-scale makers of NLT employ jigs to guide the lumber through the saw blades and maintain panel dimensions and straightness. The jigs can be made from pony walls, back and end stops, and fences. Boards are joined using a pneumatic-powered nailer, and the process is repeated until the panel is complete. Like CLT, the panel is then cut to length and fabricated to match shop drawings. Nail placement is crucial, as nails will negatively impact cutting tools, such as saws and drills.

4.2.4 MASS PLYWOOD PANEL MANUFACTURING

MPP, a recent addition to the list of mass timber products, is a veneer-based engineered wood product. The first step in its manufacture is to produce appropriately sized and graded veneer of an appropriate species. Freres Lumber Co., the only MPP manufacturer, also produces its own veneer. The MPP is then created in a two-stage process. First, billets of Structural Composite Lumber (SCL), each 1 inch thick by 4 feet wide and up to 48 feet long, are created from multiple plies of veneer. The number of plies, their grain orientation, and the grades of veneer used to create the billets vary, depending on the desired strength. In the second stage, the SCL billets are assembled into a larger and thicker

MPPs, with dimensions and strength engineered to meet the requirements of a given project.

Scarf joints (i.e., a joint connecting two billets in which the ends are beveled so that they fit over each other while maintaining a flat surface across the billets) are used to join the SCL billets, irrespective of the size of the MPP. These joints are staggered throughout the panel, so they do not create weak points. A 6-inch-thick MPP, for example, is made up of six 1-inch billets, each made of 9 plies of veneer. Thus, the total panel is made of 54 veneer plies. Throughout the manufacturing process, the entire MPP and each 1-inch SCL billet are engineered to specific strengths. Adhesive is used to bond all veneer plies in the SCL billets and to bond each SCL billet to an adjacent billet.

4.2.5 SOLID WOOD WALL (MHM)

MHM is a massive, prefabricated cross-laminated panel with layers made of rough sawn boards that are bonded with nails. This product should not be confused with one described as Nail-Laminated Timber (NLT) described above. MHM is fabricated on small-scale, turnkey three-step Hundegger production lines. The lines consist of specialized molders to produce longitudinal grooves on one side of the laminations, an automated lay-up and nailing station, and a Computer Numerical Control (CNC) finishing center. Relatively short, fluted aluminum nails that penetrate 3 layers do not interfere with cutting tools. Panels may consist of 9, 11, 13, or 15 layers (each about 16.5 mm or $^{10}/_{16}$ in).

4.2.6 DOWEL-BONDED CLT

Dowel-bonded CLT is a massive, prefabricated cross-laminated panel with layers of rough sawn boards bonded with hardwood dowels. It should not be confused with Dowel-Laminated Timber (DLT) described above. The panels are assembled in highly automated lines. The dowels are arranged in a carefully designed pattern and inserted in the lay-up by a CNC equipment. Low moisture content and tight fitting of the dowels at the time of assembly assures a durable tight connection once the dowels swell as they gain moisture in the ambient conditions. Only two commercially successful systems are known to date: 1) developed by Thoma Holz100 (or Wood 100) company in Austria; and 2) developed by Swiss industrial hardware manufacturer TechnoWood. By mid-2019, TechnoWood had installed 8 highly automated lines in Europe. Unlike other CLT products, some layers of the dowel-bonded CLT are arranged at 45 or 60 degrees to the surface layer direction.

4.3 NORTH AMERICAN MASS TIMBER PLANTS

This section provides an assessment of mass timber manufacturing capacity. Manufacturer information was collected through personal communication with manufacturers, publicly available research, information compiled by industry experts, and company profiles from websites and other published information sources. Please note that the status of manufacturing operations is constantly changing, with shifting operating schedules, some plants reaching completion, and others getting underway. The data and information that follows was current as of December 2021.

COMPLETED OUTDOOR KITCHEN STRUCTURE MADE FROM TIMBER AGE SYSTEMS CLT

Source: Timber Age Systems

CASE STUDY: TIMBER AGE SYSTEMS

SMALL TIMBER AGE SYSTEMS MAKES A BIG IMPACT

TIMBER AGE SYSTEMS manufactures Cross-Laminated Timber (CLT) panels in Durango, Colorado. As a small-scale, low-tech, low-capital manufacturer of ponderosa pine CLT panels, it stands apart from the other CLT manufacturers described in this report in its approach.

BEGINNINGS

Around 2014, during his time at a regional wood-products production facility, Kyle Hanson, Timber Age Systems founder and CEO, was challenged by US Forest Service representatives in Southwestern Colorado to find a use for the overcrowded, small-diameter ponderosa pines common to the region's forests. In 2017, Hanson consulted with his friend and eventual cofounder, Andy Hawk, and a group of trusted advisers. They identified CLT as a potential pathway to utilizing ponderosa pines, but they also realized the traditional scale of investment and extended supply chains would not be realistic for the Four Corners area. Their first step was applying for and receiving a US Forest Service Wood Innovations Grant. The key early objectives were proving that CLT panels could be made from local materials using a simple approach, and could be used in small structures.

PROVING THE CONCEPT

With grant support, Hanson and Hawk established their business in 2018. Their business model, which hasn't changed much since the beginning, is a low-tech, low-capital, fully vertically integrated operation that begins with trees from local forests and ends with CLT panels installed in local buildings.

The process starts when Timber Age Systems obtains ponderosa pine logs from local public and private landowners. According to Hawk, "All of our logs come from within about fifty miles of our shop. They come in a range of diameters, but all are cut to eleven feet in length to accommodate the dimensions of the plant's CLT panels, which are a maximum of five feet wide by ten feet long." Hawk says the decision to make 5-foot by 10-foot panels was deliberate because their research indicated that contractors wanted a panel that could be put in place without heavy equipment or cranes.

Timber Age Systems uses a semimobile band sawmill to convert the logs into 5/4-inch-thick green (undried) lumber. The lumber is sorted as it comes off the sawmill to remove the pieces with defects severe enough to make the pieces unusable. Because Timber Age Systems uses polyurethane adhesive in its CLT panels, all lumber that makes the grade is dried to about 12 percent moisture content through a combination of air-drying and solar kiln-drying. After lumber reaches the target moisture content, it is planed to 1-inch thickness and then laid up into 3-layer or 5-layer panels. The panels are formed by hand, the adhesive is applied by hand, and a com-

FINISHED TIMBER AGE SYSTEMS CLT PANEL WITH PRESS IN BACKGROUND

Source: Timber Age Systems

bination vacuum and mechanical press is used to allow the panels to cure.

Hawk says, "We run four ten-hour days per week and our four-person crew (two running the sawmill and two two laying up panels) produces an average of sixty cubic feet of panels per day." Each 3-ply panel is about 12.5 cubic feet. This means that the average production equals about 5 finished panels per day. Panels are cut to final size and machined for cutouts using hand tools. The company also has a small Computer Numerical Control machine on order that will help automate final panel finishing. All of the panel processing and lay-up is accomplished in a 1,000-square-foot shop.

MARKET ACCEPTANCE

The initial panels produced by Timber Age Systems were subjected to destructive tests at Fort Lewis College, which is also in Durango. The company has also worked with local building code officials to allow use of the panels in structures. Timber Age Systems is working toward achieving PRG 320 certification. The combined effect of these efforts is that Timber Age Systems panels have been used in a variety of small, simple outdoor structures within about 100 miles of where the trees grew and where they were milled into CLT panels. The largest building constructed to this point is a 16- by 24-foot outdoor kitchen. In another recent project, the company's panels were used in a 10- by 20-foot unattached outdoor storage shed. One objective of the shed project was to show local contractors how quickly CLT structures can be assembled. Hawk says, "The panels were flat-packed on a truck at their shop, delivered to the jobsite, and the building was erected—all in a single day."

Perhaps the most important aspect of Timber Age Systems market acceptance efforts is a strategic alliance with Clark & Chapin, an architecture studio in Cortez, Colorado. Clark & Chapin is working with Timber Age Systems to design small modular homes that can be constructed from its CLT products. According to Hawk, "The goal of the current small demonstration-type building projects combined with Clark & Chapin's efforts is that soon we'll be using our CLT panels as the structural components in small modular housing applications in Southwestern Colorado."

BUSINESS GROWTH

Hawk says the company has enough projects in the pipeline to more than stay busy, but it is planning for growth. Timber Age Systems was recently awarded a grant from the Colorado Department of Public Health & Environment through the Recycling Resources Economic Opportunity program. Because much of the raw material used by Timber Age Systems would otherwise be landfilled, it received the grant to help expand the business, and in the process, continue to divert material from landfills. Timber Age Systems plans to expand its shop to more than 3,000 square feet, allowing daily production to triple. Construction likely will begin soon.

Timber Age Systems also has long-term plans to replicate its business model in several Western US locations. That's because Hawk and Hanson believe the best recipe for growing low-capital, low-tech CLT manufacturing is locating the business near the resource (e.g., underutilized ponderosa pine trees) and then putting the materials to use in buildings less than 100 miles from where the trees were grown. They are doing that instead of scaling up the size of the existing manufacturing operation and shipping raw materials and finished products long distances.

KEYS TO SUCCESS

Reflecting on Timber Age Systems' business development, Hawk says the key to success so far has been the story the company has to tell. The business provides a solution to forest health issues by using trees that are otherwise at risk of wildfire; using an interesting, new-to-North America technology; and keeping everything local—all while using a low-capital and low-technology approach. Hawk's number one recommendation for others thinking of starting a similar business is to take the risk of developing sample structures because that allows local contractors, code officials, and potential building occupants to better understand mass timber, ultimately leading to more business. ◐

COMPANY	LOCATION	STATUS	ESTIMATED MAXIMUM ANNUAL PRODUCTION CAPACITY (M3/YEAR)
DR Johnson	Riddle, OR, United States	Operating	
Element5 #1	Ripon, QC, Canada	Operating	
Element5 #2	St. Thomas, ON, Canada	Operating	
Freres	Lyons, OR, United States	Operating	
Kalesnikoff	South Slocan, BC, Canada	Operating	
Mercer International	Spokane, WA, United States	Idled	
Nordic Structures	Chibougamau, QC, Canada	Operating	
Smartlam North America	Dothan, AL, United States	Operating	
Smartlam North America	Columbia Falls, MT, United States	Operating	
Sterling Lumber	Lufkin, TX, United States	Operating	
Sterling Lumber	Phoenix, IL, United States	Operating	
StructureCraft	Abbotsford, BC, Canada	Operating	
Structurlam	Okanagan Falls, BC, Canada	Operating	
Structurlam	Conway, AR, United States	Operating	
Texas CLT	Magnolia, AR, United States	Operating	
Vaagen Timbers	Colville, WA, United States	Operating	
Total			**1,665,000**

TABLE 4.1: CURRENTLY OPERATING NORTH AMERICAN MTP PLANTS

4.3.1 NORTH AMERICAN MASS TIMBER PLANTS' CAPACITY AND OPERATIONAL STATUS

In recent years, the North American mass timber manufacturing industry steadily increased year-over-year in terms of the number of plants and their production capacity. In 2021, however, the industry's estimated capacity contracted for the first time. As shown in **Table 4.1,** at the time of this writing (late 2021), 11 companies were operating 15 facilities in North America with a combined estimated annual production capacity of 1.6 mil-lion cubic meters per year. That compares to 12 companies in prior edition of this report who were operating 15 facilities with an estimated combined annual production capacity of 1.67 million cubic meters. Applying the rule of thumb of a 65 percent practical production factor to the maximum capacity results in an estimated practical annual production capacity of about 1.0 million cubic meters per year. A key change is that the Katerra Inc. facility went into bankruptcy before being purchased by Mercer International. The facility is idled as the new owners chart a course forward. Structurlam

FIGURE 4.2: LOCATION OF
NORTH AMERICAN MASS TIMBER
MANUFACTURING FACILITIES

Source: The Beck Group

OPERATING	UNDER CONSTRUCTION	PLANNED	CANCELED/ UNCERTAIN	SMALL SCALE	TOTAL
15	0	3	1	2	21

TABLE 4.2: SUMMARY OF NORTH AMERICAN MASS TIMBER PLANTS AS OF EARLY 2022

opened a new facility in Conway, Arkansas, offsetting the temporary loss of the Katerra facility.

At the time of this writing (late 2021), no new publicly announced mass timber plants were under construction in North America. Unchanged from last year, Stoltze Timber Systems is planning a phased approach to developing a US mass timber business based in Columbia Falls, Montana. Also unchanged from last year is that many industry observers believe Binderholz Group will eventually develop mass timber manufacturing capacity at either or both of two Southern Yellow Pine (SYP) sawmills the company purchased in 2020. One is in Florida and one is in South Carolina.

Figure 4.2 shows the location of all plants. Some are clustered in the Western regions of the US and Canada where there is a mix of available species, including Douglas fir, Western larch, hemlock, spruce, and various true firs. The rest are in Eastern Canada, where the available species are spruce, pine, and fir; and the US South, where the available species is SYP, a mix of longleaf, loblolly, slash, and shortleaf pines. No mass timber manufacturing facility has been developed in California, despite its large timber resource and its significant sawmilling industry. It is also one the largest building construction markets in North America. Euclid Timber Frames and Timber Age Systems are included for the first time this year. Both are small-scale mass timber manufacturers whose businesses are featured as case studies in this report.

The 3 plants identified as being in the planning stage in late 2020 remained in that status as of late 2021. Thus, in total, there are 21 known operating or planned mass timber manufacturing plants in North America, as summarized in **Table 4.2**.

4.4 MASS TIMBER MANUFACTURERS: COMPANY AND FACILITY DETAILS

The companies that are entering the mass timber market have diverse experience levels and strategic orientations. In North America, some firms are vertically integrated on the supply side, with sawmills and/or glulam manufacturing plants located near their panel manufacturing operations. Others are vertically integrated on the building and development end of the supply chain. Still others are stand-alone businesses. **Table 4.3** captures some of the diversity among current manufacturers by illustrating the products they offer, the status of their design guides, their brand names, etc.

4.5 NORTH AMERICAN MASS TIMBER MANUFACTURER SERVICES

Mass timber is distinct from most other wood building materials because its manufacturers tend to work closely with architects and engineers during building design regarding product specifications such as size, thickness, strength, and appearance. An important, but frequently overlooked, section of the mass timber supply

COMPANY	WEBSITE	PANEL BRAND NAME	DESIGN GUIDE	PRODUCTS	SPECIES	PANEL TYPES	PANEL THICKNESS	MAX WIDTH	MAX LENGTH	ENVIRONMENTAL CERTIFICATION
DR Johnson	https://www.drjwoodinnovations.com/	n/a	Yes	CLT, GLT, Lumber, Timbers, EWP	DF-L	A, I	3, 5, or 7 ply (4.125", 6.875", & 9.625" respectively)	10'	41.5'	FSC, GreenGuard glue specification
Element5 #1	https://elementfive.co/	E5 CLT & E5 Nano CLT	No	CLT, GLT, BOXX Panels	SPF	A, I	up to 15"	11.48'	52.5'	FSC
Freres	https://frereslumber.com/	MPP	Yes	MPP, Timbers, Plywood, Veneer	DF	A, I, M	up to 24"	11.83'	48'	American Tree Farm System
Kalesnikoff	https://www.kalesnikoff.com/	n/a	Yes	CLT, GLT, GLT Panels, Japan Zairai, Lumber	SPF, DF-L, Hemlock	A, I, M	3 to 11 ply (2.00" to 15.15")	11.48'	60'	Publicly available forest stewardship plan
Nordic Structures	https://www.nordic.ca/fr/accueil	X-Lam	Yes	CLT, GLT, GLT Panels, I-Joists	SPF (90% black spruce)	A, I	3, 5, 7, or 9 ply (3.5" up to 10.5")	8.85'	64'	FSC
Smartlam North America	https://www.smartlam.com/	SmartShaft	Yes	CLT, GLT, Elevator & Stairwell Shafts	DF-L, SPF, Hemlock, SYP	A, I, M	3, 4, 5, 7, or 9 (4.13", 5.50", 6.88", 9.63", and 12.38")	**10', 11'	**52', 51'	FSC, SFI
Sterling Solutions	https://www.sterlingsolutions.com/	Terralam	No	CLT, Lumber	SYP	M	3, 5, or 7 ply	8'	18'	No
StructureCraft	https://structurecraft.com/	DowelLam – DLT	Yes	DLT	SPF, DF, Hemlock, Sitka Spruce, Western Red Cedar, Yellow Cedar	A, I	4" up to 12.25"	12'	60.5'	FSC, PEFC
Structurlam	https://www.structurlam.com/	CrossLam	Yes	CLT, GLT	*SPF, DF-L, SYP	A, I, M	3.00" to 12.38"	10'	***40', 60'	FSC, GreenGuard glue specification
Texas CLT	http://texasclt.com/	Unknown	Unknown	CLT	SYP	A, I, M	Unknown	Unknown	Unknown	Unknown
Vaagen Timbers	https://vaagentimbers.com/	n/a	Yes	CLT, GLT, GLT Panels	SPF, DF-L	A, I	4.13" to 9.63"	4'	60'	FSC

TABLE 4.3: SUMMARY OF NORTH AMERICAN MASS TIMBER MANUFACTURERS' PRODUCT INFORMATION

*The plant in Conway uses SYP.

**The first number for width and length refers to the Columbia Falls, MT facility. The second refers to the Dothan, AL facility.

***The Conway plant can produce panels of 12' width and 60' length.

chain is the additional support services that mass manufacturers can provide their customers. The following bullet list briefly describes a number of these services. Note, however, that this is a rapidly evolving portion of the supply chain, as companies that provide those support services are emerging.

It remains to be seen which will become dominant: the model adopted by pioneering North American mass timber manufacturers—providing a one-stop, turnkey solution for their clients; or the more recent move to specialization—acting as "middlemen" among manufacturers, architects/designers, construction firms, and developers. Or there might always be a mix of both.

The following lists a variety of "details" required to move building projects from concept to reality.

4.5.1 ARCHITECTURAL DESIGN AND PROJECT SUPPORT

Design assist: Mass timber manufacturers assist architects with their design, including how best to incorporate mass timber into their building.

Engineering services: Many manufacturers employ engineers who help building designers review structural, mechanical, electrical, seismic, acoustic, fire, and other aspects of a building as they relate to the properties of mass timber products.

Modeling work: Most manufacturers assist in an array of construction documentation. Computer-aided design (CAD) services (e.g., BIM, SolidWorks, CATIA, cadwork, AutoCAD) has played a significant role in panelizing projects and identifying building assemblies. Using these tools, manufacturers can import engineering documen-

tation into CAD programs and develop robust 3D models of the project, making mass timber part of the building's structure.

Global perspective: The segment producing structural mass timber panels, often integrates architectural and engineering design, modeling, project management and construction as parts of an integrated package resulting in a building shell. Assistance in all these functions may also be offered to clients selecting external designers.

4.5.2 MANUFACTURING AND MATERIAL SUPPLY

Panel manufacturing: The manufacture of various panels at a production facility includes finger jointing lumber into mass timber panel layers (i.e., lamellas), molding/planing or surfacing the lumber, and pressing panels to the desired thicknesses, widths, and lengths.

Panel milling and finishing: This process includes additional manufacturing or CNC milling of panels to shop-specific drawings, and any architectural- or industrial-grade sanding, coating, and visual finishes. Many manufacturers list architectural and industrial finishes and can accommodate special requests for exposed elements. Some independently owned companies unrelated to mass timber manufacturers offer secondary manufacturing (CNC milling, finishing) of panels, glulam, and timbers.

Supplying connectors/hardware/fasteners: If manufacturers do not produce their own connectors and the other hardware required in mass timber buildings, they may source them elsewhere. Most manufacturing firms will provide this service.

4.5.3 CONSTRUCTION AND INSTALLATION SUPPORT

Logistics planning: Several manufacturers offer services that help with construction logistics, including just-in-time delivery of construction panels and sequencing the panel installation.

On-site: Speed and ease of installation are hallmarks of mass timber panels and key reasons for the industry's success. Because mass timber panel installation and construction are new to many building contractors, several manufacturers with construction experience provide on-site support.

4.5.4 OTHER MISCELLANEOUS SERVICES

Consulting services: Many mass timber manufacturers offer consulting services on an hourly basis. If the project requires more support to assess the practicality of mass timber elements, these companies can provide consultants during the design phase.

Steel fabrication: A variety of steel applications may be used in the construction of mass timber buildings. Some mass timber manufacturers offer in-house steel fabrication.

Renovation services and/or interior design options: Some building designs call for a complete package that includes kitchen, baths, appliances, and design elements. Some manufacturers offer such complete packages.

Environmental protection services: These focus on consultation and industrial matting, using CLT to protect specific areas from soil compaction and the impacts of heavy machinery.

Other: Most manufacturers offer shipping as a part of the package, as well as identifying any special requirements.

The future of building has begun.

We are revolutionizing the construction industry with a visionary way of building. The CREE timber-hybrid system exceeds current environmental standards, uses resources efficiently and offers great design flexibility.

We are international construction collective for regenerative building solutions. With our unique timber-hybrid system, our partners build smartly and sustainably around the globe. Together, we are revolutionizing the construction industry.

creebuildings.com

ICLT STRUCTURE DURING CONSTRUCTION
Source: Euclid Timber Frame Systems

CASE STUDY: EUCLID TIMBER FRAMES

EUCLID TIMBER FRAMES UNLOCKS INTERLOCKING CLT MARKET

EUCLID TIMBER FRAMES, LC (EuclidTF.com) is a custom designer and builder of timber frame structures based in Heber City, Utah. Euclid has been in business since 1987. In 2007, it developed an interlocking Cross-Laminated Timber product (iCLT), a no-glue panel built from beetle-kill wood and used in wall and roof systems. It is unusual among mass timber products because iCLT uses no adhesives or mechanical fasteners to join layers. The minor air voids assist in improving the insulating values of the thermal mass. Euclid Timber Frames designed iCLT wall components around what could be produced with their existing equipment to avoid having to make

a large capital investment in what, at the time, was an unknown product. This case study describes Euclid's approach to unlocking the interlocking CLT market.

WHAT IS ICLT?

iCLT is a prefabricated wooden panel made from 2 to 5 layers. Unlike other CLT products, in which the layers are typically made from lumber that is 1 to 1.5 inches thick, iCLT is made from lumber pieces that are 2¼ inches thick and 6 inches or 8 inches wide. Thus, each individual layer in the panel is much thicker than in typical CLT. Euclid worked with Brigham Young University's structural engineering department, Uni-

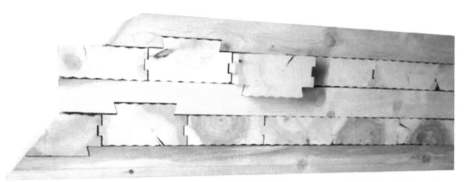

CLOSE-UP PHOTO OF ICLT
Source: itac.cap.utah.edu

LOGS HARVESTED FOR MANUFACTURING EUCLID TIMBER FRAMES' ICLT PRODUCT
Source: Euclid Timber Frame Systems

A COMBINATION TIMBER FRAME AND ICLT STRUCTURE DURING CONSTRUCTION
Source: Euclid Timber Frame Systems

versity of Utah–Integrated Technology in Architecture Collaborative (ITAC), the University of Idaho, and others to gain approval for the product's use in structural applications. Panels are designed and produced per the requirements of each structure, but the basic parameters are that panels range between 6 inches and 15 inches thick and up to 10 feet wide by 25 feet long. Testing completed at Brigham Young University found that iCLT is stronger than conventional framing and insulated concrete form walls.

HOW IS ICLT MADE?

The most unusual aspect of iCLT is that no mechanical fasteners or adhesives are used to hold the layers together. Instead, wood joinery techniques (i.e., tongue and groove, and dovetails) secure the alternating layers to each other (see photo). Euclid Timber Frames also is the first mass timber company in the US to employ a Hundegger CNC/joinery saw. Hundegger is a German company that manufactures various Computer Numerical Control (CNC) wood joinery machines. Euclid uses the K2i to process and cut joinery in the premolded feedstock boards used in fabricating the panels.

WHERE DOES THE LUMBER COME FROM?

Euclid Timber Frames harvests standing dead spruce and pine trees from the mountains surrounding Heber

FINISHED STRUCTURE ERECTED BY EUCLID TIMBER FRAMES WITH ICLT

Source: Euclid Timber Frame Systems

City, Utah. The logs are all brought to Euclid's sawmill and fabrication shop. Most of the harvested logs are converted into lumber that is 3 inches thick by 6 inches or 8 inches wide to be used as the feedstock layers for iCLT. Because the harvested lumber is standing dead in Utah's dry climate, the beetle-kill wood is air-dried to a sub-15 percent moisture content before it is machined into tongue-and-groove and dovetailed pieces.

MARKET ACCEPTANCE

Euclid Timber Frames has commercialized iCLT use in the United States, but the majority of its work continues to be traditional mortise-and-tenon solid sawn timber structures. Since building the first iCLT structure in 2011, Euclid Timber Frames has used this system in about 25 projects, including high-end residential structures and barns.

BUSINESS GROWTH

As the team at Euclid Timber Frames continues to utilize the special wood, the iCLT product keeps evolving as the team tweaks details in the number of panel layers and in the design of the joinery used to interlock the layers. For this reason, the company has kept the technology in-house to date. That strategy also fits with the simple word-of-mouth approach used to market the use of iCLT in structures. Euclid Timber Frames has, however, been able to maintain existing products while growing a sustainable iCLT business.

KEYS TO SUCCESS

At Euclid Timber Frames, the key to success has been the employees and the customers. Euclid's customers are its biggest fans. ◐

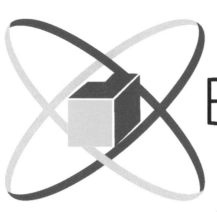

CHAPTER 5: DESIGNERS & SPECIFIERS

IMPACTS OF THE MASS TIMBER EFFECT

- Carbon neutrality by 2030 is an important goal, but the building industry can and should go further. By 2040, it can store more carbon than it emits, if mass timber market saturation is achieved.

- Choosing sustainably harvested wood as a primary structural material significantly contributes to turning a building into a carbon store.

- Of the main structural material choices for buildings, wood is the most widely available bio-based option.

- Quantifying the embodied carbon of wood products is complex, and the effort is in a nascent and rapidly developing research phase.

- A Life Cycle Analysis (LCA) for wood products assumes that the impact of forestry on emissions, sequestration, and stores of forest carbon in North America is neutral, because, overall, the growth of timber across the continent exceeds the removal.

- Forestry practices matter greatly in calculating the carbon storage potential of wood, but we do not yet have widely accepted methods to accurately measure or regulate different approaches to forest management on forest carbon pools.

- There are multiple ways of measuring a building's embodied carbon through an LCA. Designers may choose to exclude wood decomposition and presume material reuse in their carbon profiles to better understand short-term (2030) impacts.

What is the construction industry's appetite for innovation? The US Green Building Council (USGBC) considers about 5 percent of the industry to be innovators, 20 percent to be leaders, 70 percent to be followers of current codes, and 5 percent to be lawbreakers (who do not follow codes). The 25 percent who are leaders and innovators look for ways to build modern structures focused on sustainability, efficiency, and a reduced carbon footprint. Over time, as we have seen with green building certifications and their resultant effect on building codes, it is likely that these industry leaders will pull the entire building construction industry in that direction.

Mass timber is promising as an environmental solution, but it is also a disruptive technology with respect to building construction. The implications of increased off-site fabrication and new, highly collaborative construction approaches are allowing project teams to glimpse a future with increasingly higher levels of control over materials procurement and craftsmanship. As such, many designers will find that the information in Chapter 6 is equally relevant to them as teams become more integrated, optimizing the design, schedule, and costs together in real time.

This chapter also covers how to approach designing and coordinating a mass timber project from the design team perspective, from systems choices to best detailing practices to building code paths.

5.1 CARBON CONSIDERATIONS

Many designers and building owners are drawn to mass timber for its environmental credentials. A rapidly developing area of research seeks to answer their questions about how to quantify and maximize the benefits of this choice. Given that Architecture 2030, a nonprofit organization whose environmen-

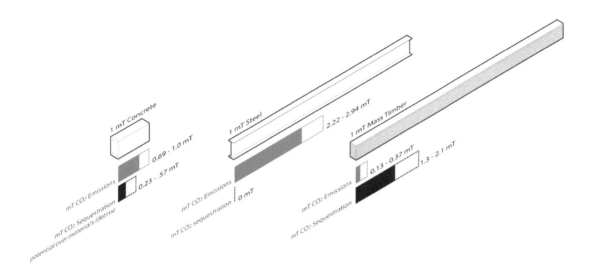

FIGURE 5.1 EMBODIED AND BIOGENIC CARBON IN COMMON STRUCTURAL MATERIALS

Source: Timber City Research Initiative, Gray Organschi Architecture, timbercity.org

tal goals have been adopted by the American Institute of Architects (AIA) in the form of the AIA 2030 Challenge, has identified a time frame of 10 years[1] to reach net zero emissions in the building industry to curb catastrophic climate change, getting it right is crucial. This section outlines the tools and techniques for selecting and measuring the carbon impacts of mass timber in building projects. We also discuss how choosing to use mass timber, especially at scale as the market sector grows, also ultimately impacts land-use and forestry practices.

5.1.1 ENVIRONMENTAL IMPACT OF BUILDING MATERIALS

Analyzing and comparing the environmental impacts of building materials is complicated but crucial to achieving the industry's carbon goals. Embodied carbon and biogenic carbon, as defined below, are two important concepts that underlie such an analysis. To track progress, designers can use industry-developed tools that assist with environmentally conscious decision-making processes that include LCAs and Environmental Product Declarations (EPD). Several certification programs are designed to help building projects measure, meet, and promote their goals.

Embodied Carbon

Most processes involved in the extraction, manufacture, transport, and installation of building products rely on fossil fuels. The total amount of carbon emitted by a given product during this process is the embodied carbon of that product. Wood products have much lower embodied fossil energy content than concrete or steel because they require significantly less energy to produce (see **Figure 5.1**). We frequently compare wood with these two other materials because the structural

1   *Architect Magazine*, The Carbon Issue, January 2020, guest edited by Architecture 2030.

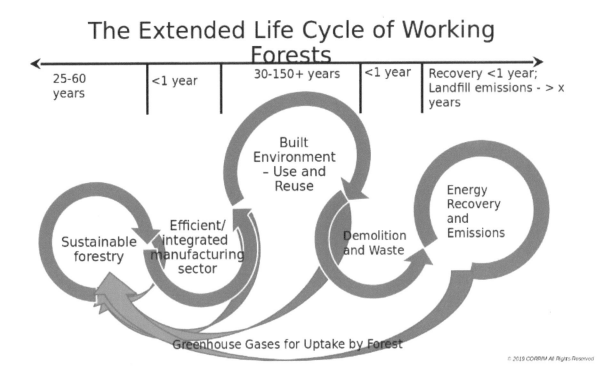

The Extended Life Cycle of Working Forests

| 25-60 years | <1 year | 30-150+ years | <1 year | Recovery <1 year; Landfill emissions - > x years |

Built Environment – Use and Reuse

Sustainable forestry

Efficient/integrated manufacturing sector

Demolition and Waste

Energy Recovery and Emissions

Greenhouse Gases for Uptake by Forest

FIGURE 5.2: EXTENDED LIFE CYCLES OF WORKING FORESTS

Source: Reprinted with permission, Elaine Oneil, Consortium for Research on Renewable Industrial Materials (CORRIM), www.corrim.org

system of a building comprises up to 80 percent of its entire embodied carbon. Wood is an effective replacement of these widely used, high-embodied-energy structural materials. In fact, wood products are often produced substantially with renewable energy, including the combustion of manufacturing by-products for power generation.

Architecture 2030 has determined that "embodied carbon will be responsible for almost half of total new construction emissions between now and 2050."[2] The crucial benefits of reduced embodied carbon are achieved while a building is under construction. Bio-based products also stand apart from other materials in that they

store carbon as well, potentially offsetting carbon impacts from other materials.

Biogenic carbon

"Biogenic carbon refers to carbon that is sequestered from the atmosphere during biomass growth and may be released back to the atmosphere later due to combustion of the biomass or decomposition."[3] One cubic meter of wood stores approximately one ton of carbon dioxide.

Wood, as a building material, provides long-term biogenic carbon storage. As illustrated in **Figure 5.2**, carbon storage in long-lived wood products

---

2   Architecture2030," https://architecture2030.org/new-buildings-embodied/.

3   https://www.sciencedirect.com/topics/engineering/biogenic-carbon

can extend the carbon cycle. Constructing buildings with wood products increases the length of time that carbon is kept in storage, avoiding release into the atmosphere through forest decay or fire.

Biogenic carbon eventually returns to the atmosphere through decomposition or incineration, and that may be acknowledged through a complete LCA that illuminates long-term impacts (at least 100 years). However, while end-of-life considerations are crucially important to a circular economy (see Chapter 8), most buildings built today will remain standing long after global carbon reduction timelines have passed. When calculating the total life cycle of a wood product, project teams should consider whether to include or exclude biogenic carbon, acknowledging the eventual return of the carbon to the atmosphere—or not. Total decomposition may be an unlikely occurrence, based on the expectation that structural wood will be reused or encapsulated in a landfill, rather than incinerated or mulched. Additionally, climate crisis goals should be taken into consideration.

Absorbing as much atmospheric carbon as possible in the next 10 to 30 years is a global priority to avoid irreversible climate change. The World Green Building Council (WorldGBC) stresses the importance of reducing "up-front" or embodied carbon in their 2019 report, "Bringing Embodied Carbon Upfront."[4] The report states: "To achieve our vision, we must take urgent action to tackle upfront carbon while designing with whole life carbon in mind." It can be argued that the embodied carbon stored today is more important than accounting for unknowns in deconstruction approaches, fire, or decay past that crucial timeline. Considering the urgent 10-year timeline we face globally to eliminate emissions in the industry, project teams may choose to emphasize the short-term effects of using wood products.

Buildings as Carbon Banks

On a global scale, the building industry stands out as having the potential to turn from being the largest contributor of global carbon emissions to becoming a massive atmospheric absorber. Buildings are long-lived and profoundly materials intensive, and, therefore, present an opportunity to become carbon storage devices, or carbon banks. To achieve this, the industry must use as many biogenic materials as possible in every building.

The longer a biogenic, carbon-rich building remains standing, the more effective a carbon store it is. And, because mass timber components have a high potential to retain value after the life of a building, markets for reuse will likely develop[5] for mass timber, prolonging use and further delaying decomposition. In fact, decomposition is an unlikely outcome. A worst-case scenario would send these valuable building components to a landfill, where LCAs typically assume the wood will decompose. In fact, the EPA estimates that 88 percent of the carbon in landfilled wood is permanently sequestered, and the remaining 12 percent is captured for reuse as fuel, offsetting fossil-sourced fuel usage.[6]

Life Cycle Analyses

LCAs are a process for documenting embodied carbon in building materials and comparing simi-

---

4   https://www.worldgbc.org/sites/default/files/WorldGBC_Bringing_Embodied_Carbon_Upfront.pdf
5   https://corrim.org/carbon-economy-workshop/
6   *Documentation for Greenhouse Gas Emission and Energy Factors Used in the Waste Reduction Model (WARM) (2019).*

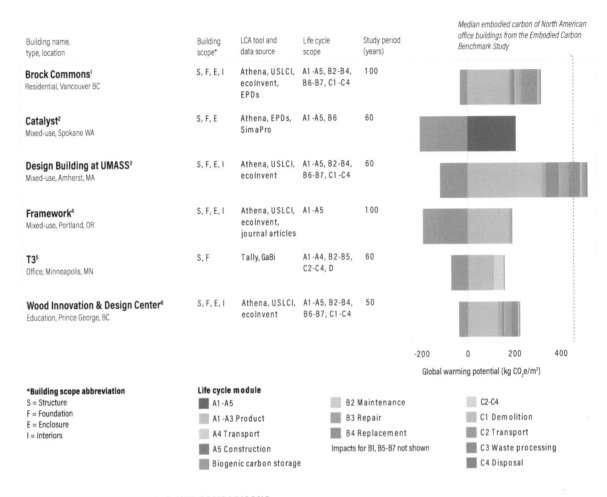

Building name, type, location	Building scope*	LCA tool and data source	Life cycle scope	Study period (years)	
Brock Commons[1] Residential, Vancouver BC	S, F, E, I	Athena, USLCI, ecoInvent, EPDs	A1-A5, B2-B4, B6-B7, C1-C4	100	
Catalyst[2] Mixed-use, Spokane WA	S, F, E	Athena, EPDs, SimaPro	A1-A5, B6	60	
Design Building at UMASS[3] Mixed-use, Amherst, MA	S, F, E, I	Athena, USLCI, ecoInvent	A1-A5, B2-B4, B6-B7, C1-C4	60	
Framework[4] Mixed-use, Portland, OR	S, F, E, I	Athena, USLCI, ecoInvent, journal articles	A1-A5	100	
T3[5] Office, Minneapolis, MN	S, F	Tally, GaBi	A1-A4, B2-B5, C2-C4, D	60	
Wood Innovation & Design Center[6] Education, Prince George, BC	S, F, E, I	Athena, USLCI, ecoInvent	A1-A5, B2-B4, B6-B7, C1-C4	50	

Median embodied carbon of North American office buildings from the Embodied Carbon Benchmark Study

Global warming potential (kg CO_2e/m^2)

-200   0   200   400

*Building scope abbreviation
S = Structure
F = Foundation
E = Enclosure
I = Interiors

Life cycle module
- A1-A5
- A1-A3 Product
- A4 Transport
- A5 Construction
- Biogenic carbon storage

- B2 Maintenance
- B3 Repair
- B4 Replacement
- Impacts for B1, B5-B7 not shown

- C2-C4
- C1 Demolition
- C2 Transport
- C3 Waste processing
- C4 Disposal

FIGURE 5.3: MASS TIMBER BUILDING GWP COMPARISONS

Several LCA studies of mass timber buildings in North America show that mass timber buildings: (1) can have low embodied carbon compared to a benchmark value, which in this figure is represented by the vertical red dotted line; (2) and can have a significant potential to store biogenic carbon. Note that this figure does not aim to compare the buildings, but instead shows the general range in global warming potential results and the variation in LCA methods and tools. Direct comparison of environmental impacts between projects is challenging due to variation in model scope, building elements, background data, and underlying methods.

Source: Carbon Leadership Forum

1   M. Bowick, *Brock Commons Tallwood House, University of British Columbia: An Environmental Building Declaration according to EN 15978 Standard* (Athena Sustainable Materials Institute, 2018), http://www.athenasmi.org/wp-content/uploads/2018/08/Tallwood_House_Environmental_Declaration_ 20180608.pdf.

2   M. Huang, C. X. Chen, F. Pierobon, I. Ganguly, and K. Simonen, *Life Cycle Assessment of Katerra's Cross-Laminated Timber (CLT) and Catalyst Building: Final Report* (Carbon Leadership Forum, 2019), https://carbonleadershipforum.org/download/5173/.

3   M. Bowick, *Design Building, University of Massachusetts, Amherst: An Environmental Building Declaration according to EN 15978 Standard* (Athena Sustainable Materials Institute, 2017), http://www.athenasmi.org/wp-content/uploads/2017/04/ UMass_Environmental_Declaration_ 31_January_ 2017.pdf.

4   S. Liang, S. Gu, R. Bergman, and S. Kelley, *Comparative Life-Cycle Assessment of a Mass Timber Building and Concrete Alternative* (USDA Forest Products Lab, 2020), https://www.fpl.fs.fed.us/documnts/pdf2020/fpl_2020_liang001.pdf.

5   Based on Tally output files received from Magnusson Klemencic Associates (MKA), March 2021.

6   M. Bowick, *Design Building, University of Massachusetts, Amherst: An Environmental Building Declaration according to EN 15978 Standard* (Athena Sustainable Materials Institute, 2015), http://www.athenasmi.org/wp-content/uploads/2015/06/ WIDC_Environmental_Declaration_final.pdf.

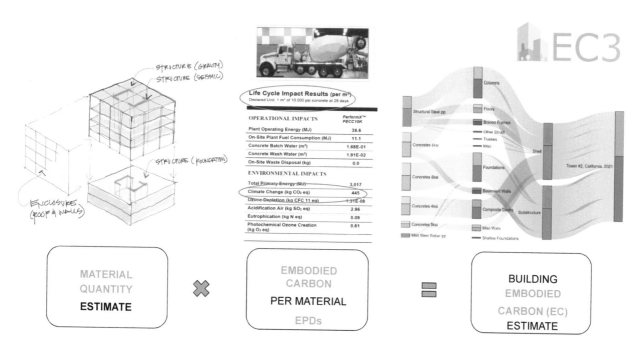

Source: Embodied Carbon in Construction Calculator Carbon Leadership Forum

lar products. An LCA might focus on a single component or product, or it might capture an entire building project. As discussed in the topics above, when calculating the LCA of a timber building, biogenic carbon can be approached by considering a decomposition or industrial reuse cycle.

The Consortium for Research on Renewable Industrial Materials (CORRIM) is a leading resource on LCAs for a variety of wood products. Embodied carbon and Global Warming Potential (GWP) have been researched and calculated for several North American mass timber products, yielding a range of results because of variations in wood sourcing and manufacturing processes. As more research and data are available, the current, educated assumption that wood products can, depending on the source, more than offset the carbon required to produce and install them will be refined.

The Carbon Leadership Forum (CLF) is widely trusted for producing best-practices Whole Building LCAs (WBLCA) for timber structures. In a study for Katerra in 2019, CLF compiled information from several mass timber buildings to compare their GWP from a WBLCA standpoint. **Figure 5.3** shows the buildings' GWP both with and without biogenic carbon included, and in relationship to similar buildings with primary structural systems of concrete or steel.

LCA tools available to designers include Tally,[7] popular for its ability to plug in to Revit; Athena; BEES (Building for Environmental and Economic Sustainability); and, more recently, EC3 (Embodied Carbon in Construction Calculator), illustrated in **Figure 5.4**. EC3 is a free, open-source LCA tool released in late 2019 and developed by a multidisciplinary team led by the CLF, and the most so-

---

7   https://kierantimberlake.com/page/tally

CrossLam CLT
Structurlam

Final Assembly: Okanagan Falls, British Columbia, Canada
Life Expectancy: 100 Years
End of Life Options: Recyclable (100%)

Ingredients:

Organic Wood: Softwood (Spruce Pine Fir); CLT - Face Bond Glue - Maple St.: Polymethylene Polyphenyl Isocyanate, Polymethylenepolyphenylene Ester, Methylenediphenyl Diisocyanate, Diphenylmethane-2.4'-Diisocyanate, Siloxanes and Silicones, Di-Me, Reaction Products with Silica, 2,2'-Methylenediphenyl Diisocyanate; Water; Face Bond Primer: Sorbitan, Monododecanoate, Poly(Oxy-1,2-Ethanediyl) Derivs.

Living Building Challenge Criteria:

SLM-0001	EXP. 01 DEC 2020
VOC Content: N/A	VOC Emissions: CDPH Compliant
Declaration Status	■ LBC Red List Free
	☐ LBC Compliant
	☐ Declared

MANUFACTURER RESPONSIBLE FOR LABEL ACCURACY
INTERNATIONAL **LIVING FUTURE** INSTITUTE™ declareproducts.com

FIGURE 5.5: ENVIRONMENTAL PRODUCT DECLARATION FOR CROSS-LAM CLT

Source: https://declare.living-future.org/

phisticated to date. Each tool will vary somewhat in end-of-life options and assumptions, and users of these tools will find that these factors contribute greatly to the output of LCAs.

Environmental Product Declarations

Reducing embodied carbon in building products reduces their GWP. Designers can reference the information for products where GWP is mea-

sured and published, along with other disclosures like toxicity or land conversion, by reviewing the product's EPD. EPDs report on five categories of environmental effects: GWP, ozone depletion potential, acidification potential, smog potential, and eutrophication potential. EPDs completed in compliance with the International Organization for Standardization (ISO) 14025 Type III are prepared and reviewed by an independent third party.

EPDs allow a specifier to compare different materials that provide the same function in a construction project. Though a manufacturer may choose to pursue EPDs specific to their products—especially if they have exceptionally good reports—general EPDs for wood products are available through the American Wood Council and the Canadian Wood Council. One of the most demanding EPD labels is "Declare" (See **Figure 5.5**); it identifies the most dangerous "red list" ingredients and clearly states when products are free of them. Four Cross-Laminated Timber (CLT) manufacturers have achieved this label for their products (listed in the Adhesives section, below).

EPDs are complex to interpret and time-consuming to track down, but they are becoming more accessible as building owners and industry professionals demand nontoxic and low-carbon materials. Some excellent and rapidly expanding resources for designers include the databases Mindful Materials[8] and Carbon Smart Materials Palette,[9] and the organizational tool EPD Quicksheet.[10]

---

8   http://www.mindfulmaterials.com/
9   https://materialspalette.org/palette/
10   https://architecture2030.org/epd-quicksheet/

FIGURE 5.6: ILFI'S ZERO CARBON CERTIFICATION
REQUIRES EMBODIED CARBON DISCLOSURES

Green Building Certification Programs

The pursuit of environmental certifications is optional for most projects, but these programs and their supporters generally believe there are financial and nonfinancial benefits. These benefits include recognition/prestige, tax incentives, reduced ongoing operating costs, faster lease-up times, increased property values, increased energy efficiency, reduced waste, and healthier, more enjoyable working/living conditions for tenants.

Options for certification programs include Leadership in Energy and Environmental Design (LEED), Green Globes, Passive Haus, and International Living Future Institute's (ILFI) suite of living building approaches. Each of these programs has different criteria for certifications. All, however, share a mission to construct buildings with reduced environmental impacts. The use of wood as a building material is generally seen as positive within the context of the evaluation processes, though they vary in how wood certifications are viewed and accepted.

Zero-carbon certifications have emerged over the last several years in response to the growing realization of the importance of neutralizing embod-

ied carbon in the building industry. Internationally, projects can register with ILFI's Zero Carbon Certification program (**Figure 5.6**), requiring that, "One hundred percent of the embodied carbon emissions impacts associated with the construction and materials of the project must be disclosed and offset."[11] The Canada Green Building Council (CaGBC) has a Zero Carbon Building (ZCB) Standard that recognizes embodied energy as well as operational energy. To date, 10 ZCB Standard projects have been completed. USGBC's LEED Zero tracks operational energy only, but LEED's newest version, 4.1, awards credits for embodied carbon accounting.

These building certification programs, where wood building products are concerned, often tie back into forest management certifications, solidifying the connection between sustainably managed forests and the utilization of wood in new and creative approaches to construction. These systems continually extend the goal of creating human habitat with an ever-smaller environmental footprint, and increasingly recognize that using wood is a significant component of that goal.

5.1.2 IMPACT OF BUILDING MARKET DEMAND ON FOREST CARBON

Many architects who choose to work with wood will be asked about forestry and logging, and that, for many, will be the first time they've had to consider from where exactly their raw building materials come. These questions tend not to come up with inorganic materials like steel and concrete, though, of course, everything comes from somewhere. The emotional connection people have with trees may be behind this investigative imperative.

---

11   https://living-future.org/zero-carbon-certification/

RENDERINGS: PORT OF PORTLAND

Source: ZGF Architects; Photo Credit: Stephen A. Miller

CASE STUDY: PORTLAND INTERNATIONAL AIRPORT MAIN TERMINAL EXPANSION

SOURCING FOR CLIMATE AND COMMUNITY RESILIENCE

STORY CREDIT: ZGF ARCHITECTS, PORT OF PORTLAND

LOCATION: PORTLAND, OREGON

COMPLETION DATE: 2025

OWNER: PORT OF PORTLAND

ARCHITECT: ZGF ARCHITECTS

STRUCTURAL ENGINEER: KPFF

CONTRACTOR: HOFFMAN SKANSKA JOINT VENTURE

MT MANUFACTURER(S): ZIP-O-LAMINATORS, CALVERT, FRERES LUMBER

MT CNC FABRICATION: TIMBERLAB

MT INSTALLATION: SWINERTON

MT TRADE PARTNER (DESIGN ASSIST AND PREFABRICATION): SWINERTON

WOOD ADVISER: SUSTAINABLE NORTHWEST AND SUSTAINABLE NORTHWEST WOOD

Source: ZGF Architects
Photo Credit: Stephen A. Miller

LUMBER MILLS: ZIP-O-LOG MILLS, MANKE LUMBER, FRANK LUMBER, KASTERS KUSTOM CUTTING, ELK CREEK FOREST PRODUCTS, HERBERT LUMBER, FRERES LUMBER

FORESTS: YAKAMA NATION, THE NATURE CONSERVANCY CLE ELUM, HYLA WOODS, SKOKOMISH INDIAN TRIBE, DONEEN FOREST, COQUILLE INDIAN TRIBE, WILLAMETTE NATIONAL FOREST

FSC CREDIT PROVIDER: ELK CREEK FOREST PRODUCTS

CERTIFICATIONS: TARGETING LEED GOLD V4

Hailed by *Travel + Leisure* magazine as "America's Best Airport," Portland International Airport plans to significantly increase its capacity over the next 2 decades to accommodate 35 million passengers annually. The ZGF-designed main terminal expansion doubles the footprint while inviting passengers and employees alike to celebrate the beauty of the Pacific Northwest.

A stunning wooden roof greets visitors as they arrive and celebrates the state's history of forest product innovation with wood sourced from landowners and mills within a 600-mile radius. The undulating canopy of Mass Plywood Panel (MPP) and glulam is penetrated by skylights over 34 Y-shaped columns that hold the 18-million-pound, 380,000-square-foot roof in place.

Through careful planning, dialogue with landowners and mills, and more than a bit of ingenuity, the project achieved something no other project at this scale has done: to track around 40 percent of the project's glulam beams and timber lattice ceiling (~1,000,000 board feet) directly back to their source. For the remaining 60 percent, the mills agreed to provide information about the larger mix of regional landowners who contributed to specific lumber batches.

Careful consideration was given to ensuring the wood was harvested from landowners using healthy

Source: ZGF Architects
Photo Credit: Stephen A. Miller

forest management practices. The team also set limits on how much wood volume could come from a single landowner while targeting forests on both the west and east sides of the Cascades, helping to guarantee that a larger diversity of landowners was represented.

More than 95 percent of the roof beams and the lattice ceiling is sustainably harvested through a mix of certified Forest Stewardship Council (FSC) lumber, sourcing from federal lands with specific stewardship contracts, and custom verification of sustainable sourcing criteria developed by the project team. These custom criteria considered whether landowners maintain diversity of species within the forest landscape, clear-cut smaller openings with

more retention, increase restrictions in riparian management zones, and manage multiage forests. FSC certification was used when the mills could not provide forest-of-origin information, while segregated wood was custom-verified to come from sustainable sources, with or without forest certification.

Paying homage to the sense of place, the regionally and sustainably sourced timber that forms the basis of the roof can be traced back to its forest of origin, honoring the small families, Pacific Northwest tribes, and other landowners who contributed to its creation. ⬤

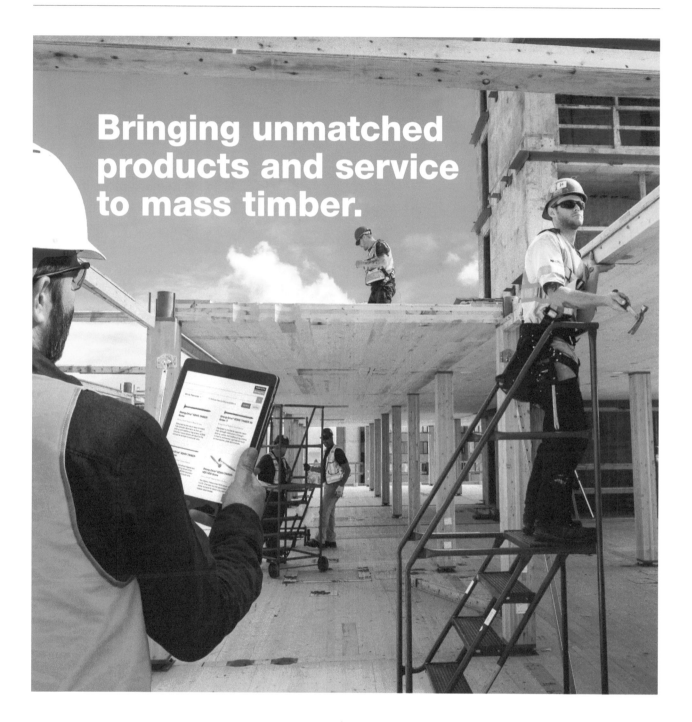

Bringing unmatched products and service to mass timber.

With over 60 years of leadership in structural engineering, Simpson Strong-Tie is now proud to offer smart solutions for mass timber. From our rigorously tested connectors and fasteners that provide design flexibility, to a nationwide supply network that delivers exactly what you need, when you need it — our products and expertise ensure that your mass timber projects are built faster, easier and stronger than ever.

To learn more, visit **go.strongtie.com/masstimber** or call (800) 999-5099.

Land Use

One of the biggest concerns in using forest-sourced products is the fear of causing deforestation or forest degradation. Forest degradation can occur when logging practices cause biodiversity loss or reduce the ecological resilience of an ecosystem. Designers should consider the sources of the fiber they specify, and they can turn to forest certifications as one way to support sustainable forest practices. Chapter 2 contains further discussion on certifications.

The biggest cause of deforestation is not fiber harvests, but agriculture and development. When land is not valued as forests, it tends to get turned into something else, all around the world. Thus the idea, counterintuitive at first but economically logical, that using forest products may contribute to an increase in lands used for forestry, and, in turn, increased carbon stores in forests.[12]

Forestry Practices

An increased demand for forest products appears to also drive more sustainable forestry practices. According to the CLF, "Transitioning construction of low to mid-rise commercial and non-residential structures to cross-laminated timber (CLT)/heavy timber construction could have a positive impact on the environment. It could also develop a new market for the smaller diameter and lower quality logs derived from forest thinning and forest health operations, thereby providing an incentive to undertake forest management activities designed to improve forest health and resiliency. Finally, the development of a cross-laminated timber industry would provide substantial economic benefits and employment opportunities for rural timber-dependent communities."[13]

5.2 ELEMENTS OF DESIGN

Wood is one of the oldest building materials. As far back as 6,000 BCE, humans made dwellings using wood. Wooden longhouses sheltering more than 20 people date to at least 4,000 BCE. To build large wooden structures, humans have long taken advantage of wood's natural strength while minimizing any weaknesses. Over the millennia, building techniques and capabilities have improved, most recently with the development of mass timber panel systems.

5.2.1 PANEL SIZE

Mass timber panels are groundbreaking in the engineered wood market because their scale necessitates prefabrication and creates the potential for their use in modular construction. To maximize the benefits of mass timber panels, a building designer must consider the panel as it relates to the building's grid system in terms of overall dimensions, as well as the number of laminations and panel thickness. Each manufacturer has different fabrication machinery and thus different size limitations. In North America, a typical panel might be around 10 feet by 40 feet nominally, with between 3 and 7 laminations. There are, however, many other options. A designer must also consider the actual, versus the nominal, dimensions when designing with mass timber panels. One of the characteristics of CLT panels is their remarkable dimensional stability, particularly in-plane. This enables manufacturers to finish the panels for construction with submillimeter precision,

---

12 Reid Miner, "Presentation to Carbon Leadership Forum" (September 9, 2020).
13 https://carbonleadershipforum.org/cross-laminated-timber-optimization/

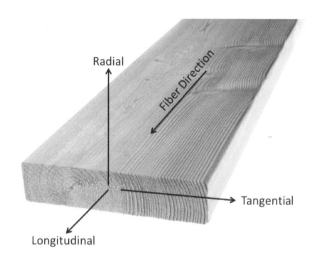

FIGURE 5.7: LUMBER STRENGTH ILLUSTRATION

removing the difference between nominal and actual panel dimensions, as discussed in Chapter 3.

Panel sizes have also been developed around transportation requirements. The transport limitations at any given building site should be considered when choosing optimum panel sizes for a project, though some may find construction efficiency justifies the high cost of transportation of oversized elements and assemblies.

5.2.2 PANEL STRENGTH

As mentioned in Chapter 1, engineered composite wood products are stronger than solid wood components of the same dimensions because of the redistribution of natural defects in the wood. Mass timber panels truly take advantage of the natural strengths of wood while minimizing its natural weaknesses. Wood is naturally much stronger in the longitudinal direction (aligned with the grain) than in the radial and tangential directions (across the grain), illustrated in **Figure 5.7**. Products like CLT and Mass Plywood Panels (MPP) take advantage of wood's longitudinal strength by alternating the grain direction in each

layer, resulting in a panel that is strong and dimensionally stable in both in-plane directions.

During the ongoing development of mass timber products, testing, including measurements of the strength of various panel styles and assemblies, has been constant. Because there are innumerable panel variables (number of layers, species of wood, lumber sizes and grades, adhesives versus fasteners), the testing has taken two approaches: (1) physically testing specific panel size/layers/species configurations, and (2) extending the physical test results to other untested size/layers/species configurations through analysis and modeling. The combination of an analytical approach and experimental testing has created a baseline understanding of the strength of mass timber products.

For detailed information on design standards for mass timber products, refer to **Table 5.2.**

5.2.3 ADHESIVES

Adhesives are used in most engineered wood products, including plywood, Laminated Veneer Lumber (LVL), glulam, CLT, and MPP. Standards have been established to ensure that these adhesives are structurally reliable and safe.

Requirements for adhesives used in glulam and CLT are very similar. Adhesives used in glulam must meet the requirements of the American National Standards Institute (ANSI) *405 Standard for Adhesives for Use in Structural Glued Laminated Lumber (ANSI 405)*. Guidance for CLT, under PRG 320, specifies that adhesives in CLT used in the US must also conform to ANSI 405, with two exceptions. First, Section 2.1.6 of ANSI 405 does not apply because it is intended to ensure glue-bond durability in exterior applications, and

CLT is not recommended for exposed exterior applications. The second exception is that for the small-scale flame test under CSA O177 (Sections 2.1.7 and 3.7 of ANSI 405), CLT must be substituted for glulam.

PRG 320 specifies that adhesives in CLT used in Canada must conform to CSA O112.10 and Sections 2.1.3, 2.1.7, 3.3, and 3.7 of ANSI 405 with the same alteration to the small-scale flame test under CSA O177 as is required in the US. In addition, for both the US and Canada, PRG 320 specifies that CLT adhesives must conform to Annex B of PRG 320, which lays out standards for testing during elevated temperatures.

In CLT, the most used adhesives are polyurethane (PUR) -based, but melamine formaldehyde resins are also used. MPPs use a phenol formaldehyde adhesive like those used in plywood and LVL. These adhesives are continually being studied and refined to be both better for the environment and to better meet strength objectives desired by the industry. Some manufacturers in Europe and in North America use urea formaldehyde (UF) or melamine urea formaldehyde (MUF) resins in their processes. Manufacturers in Japan use a variety of other cold curing adhesives, though mostly in small-scale operations.

Many mass timber products have EPDs available that demonstrate the safety of their adhesives from a health standpoint. In fact, at least three CLT manufacturers with North American availability have achieved "red-list free"[14] status (and one other is "red-list approved"[15]) by the ILFI's Declare EPD label, the most rigorous of sustainable building standards.

Bio-based adhesives are an area of interest for designers and manufacturers looking for low-toxicity and low-carbon products.

5.2.4 CONNECTORS

As mass timber construction increases, so does the need for proper fasteners and connectors. Connectors are used to join the structural components and to transfer loads throughout a building. A variety of factors must be considered when using numerous connectors in a mass timber building, including the type of joint, the materials being joined, loads carried through the joint, and aesthetics. Connectors range from nails and screws to more complicated bracket systems, and they include glued-in, or dry insert, wooden or steel rods. See **Figure 5.8** for examples. Some of these systems are proprietary, while others are traditional and widely available.

Connectors and fasteners must meet specific engineering requirements and be tested for performance. Two important requirements are shear strength and withdrawal strength. Shear strength is the ability of a material to resist forces that can cause the internal structure of the material to slide against itself (that is, fail) along a plane parallel with the direction of the force. Withdrawal strength, or withdrawal capacity, is the ability of the connector to resist forcible removal, or tear out, from its entry point. The National Design Specification (NDS) for Wood Construction provides design values for most dowel connectors, as well as for shear plates and split rings, while design values for proprietary systems are found in code evaluation reports, which the manufacturer can provide.

---

14   Structurlam, KLH, and Nordic Structures, https://declare.living-future.org/.
15   Katerra

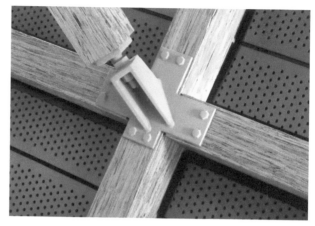

FIGURE 5.8: MASS TIMBER CONNECTOR EXAMPLES

Sources: APA, The Engineered Wood Association, Structure Craft (upper right);
Oregon Department of Forestry (lower left)

With all connectors, it is important to know where to find their applicable design values. The International Building Code (IBC) defines the structural property requirements for connectors and fasteners of wood components. Section 2302.1 lists the various sections that cover the actual stress factors required for various building applications. Sections 2304.10.1 through 2304.10.7 define the requirements for connectors and fasteners: what types of fasteners are to be used in what situations, how many, and where they should be placed.

There are two primary families of connections for wood construction: traditional joinery; and mechanical, including dowels, splines, plates, and other specialized, usually metal, components.

Joinery

Joinery uses specialized cutting techniques to form joints between wood components (mortise and tenon, dovetail, etc.). Joinery can create impressive results, both in beauty and strength. Long understood to be a time-consuming manual process that requires a significant amount of skill, the possibilities have become more accessible to

FIGURE 5.9: CNC JOINERY WITH PREFABRICATED MASS TIMBER COLUMNS AND BEAMS

Source: Emily Dawson,
Tamedia Building, Zurich, Switzerland

FIGURE 5.10: WOOD NAIL COIL AND LIGNIN WELDING
Source: LIGNOLOC®

the building market through Computer Numerical Control (CNC) technologies (see **Figure 5.9** for an example). Designs translated into a computer model to be read by the CNC operator can be unique and imaginative, or they can be optimized for material efficiency and speed—or potentially both. The intricacy of a given design will affect the time spent in cutting and assembling custom wood profiles, and that, in turn, will impact cost. Working with a fabricator early in the design process can inform the cost-effectiveness of a joinery-based design approach.

Dowels

The most common type of mechanical fastener, dowel connectors can be made from a variety of materials. Metal dowel connectors are typically steel, and they include staples, nails, screws, and bolts. Dowel connectors transfer loads well, and they are generally easy to install and cost-effective.

While wood dowels can technically provide both a mechanical connection and a lignin bond, their application is analogous to metal dowel connectors. The NDS for Wood Construction allows designers and engineers to calculate the strength properties of dowel connectors. (See also NLT and DLT in Chapter 1.) The benefits of wood doweling as a mass timber connection or fabrication approach are twofold: a higher carbon sequestration potential, and a more readily reusable or recyclable product at end of life. "All-wood" timber products that do not include added metal or adhesives generate an improved LCA profile.

Recent testing at the University of Hamburg identified the phenomenon of "lignin welding," which led to the development of wooden nails acceptable for structural applications. Subsequently, a proprietary wooden nail product (see **Figure 5.10**) made from beechwood was developed in Austria, utilizing the lignin welding effect. The German Institute for Construction Engineering (DIBt) recently issued technical approval of load-bearing timber connections using these wooden nails, noting "...(T)he large amount of heat generated

by friction when the nail is driven in at a high speed causes the lignin of the wooden nail to weld with the surrounding wood to form a substance-to-substance bond."[16]

Splines

Spline connections combine joinery concepts and dowel connectors to structurally join large mass timber panels with smaller-scale Engineered Wood Products (EWP). A typical spline connection involves routing the connecting edges of two mass timber panels with a shallow groove, laying joinery boards within the groove, and fixing them in place with nails or screws (see **Figure 5.11**).

FIGURE 5.11: SPLINE CONNECTION MATERIAL
EXAMPLES: PLYWOOD AND JOINERY BOARD

Source (Structurlam CLT and plywood splines): Emily Dawson
Source (KLH 1-inch joinery board): Scott Noble

---

16 Beck, "LIGNOLOC®," press release, September 23, 2020.

FIGURE 5.12: SHEAR PLATE CONNECTOR

Source: Portland Bolt & Manufacturing Co.

Plates

Metal connector plates were developed to help join trusses for floors and roofs. These plates are usually made from sheets of galvanized steel and are die-punched to create teeth that protrude from the underside of the plate's face. This type of toothed metal connector plate is generally not suitable for mass timber applications.

Shear Connectors

Shear connectors, or bearing connectors, include shear plates, toothed shear plates, and split rings.

These connectors are designed to help wooden components handle heavier loads. Shear plates, or timber washers, are iron discs with a shallow rim on one side and flat surface on the other (see **Figure 5.12**). This connection disperses pressure from a load across the larger radius of the plate. By contrast, a bolt spreads pressure across a smaller area. Shear plates, therefore, can handle heavier loads than bolts. Split rings are like shear plates in both form and function, but they are not as heavy-duty as the discs.

FIGURE 5.13 OFF-THE-SHELF STRUCTURAL METAL CAST COLUMN CONNECTIONS
TIMBER END CONNECTORS™, UMASS AMHERST INTEGRATED DESIGN BUILDING

Source: Cast Connex®; Photo Credit: Alex Schreyer

Structural Metal Castings

The free-form capability of the casting manu-facturing process is ideally suited to address a variety of connection geometries with artistic cre-ativity and structural integrity. Structural metal castings can transfer tension, compression, shear, and other loads, as well as offer increased ductil-ity for structural systems that are meant to resist seismic motions. Pre-engineered, standardized castings are available off-the-shelf to suit an array of member sizes. Custom-designed cast connec-tions can satisfy specific project objectives and constraints for one-off and repetitive applications (see **Figure 5.13**).

Proprietary Connector Systems

Proprietary connector systems are numerous and vary significantly in appearance, capacity, and ap-plication. These systems range from self-tapping screws with proprietary head patterns to one-off, custom-created connectors that weigh hundreds or thousands of pounds.

Self-tapping screws are one of the most widely used fasteners in mass timber projects. Propri-etary bracket systems are also commonly used to connect beams, posts, and panels. Proprietary systems can be created for a variety of reasons. Some are intended to overcome limitations or weaknesses in existing systems or components

Effective Char Rates and Char Layer Depths
(for β_n = 1.5 inches/hour)

Required Fire Resistance (hr)	Effective Char Rate, β_{eff} (in/hr)	Effective Char Layer Depth, a_{char} (in)
1-Hour	1.8	1.8
1½-Hour	1.67	2.5
2-Hour	1.58	3.2

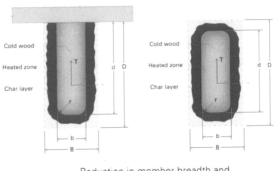

Reduction in member breadth and depth over time, t

FIGURE 5.14: CALCULATING THE FIRE RESISTANCE OF EXPOSED WOOD MEMBERS

Source: American Wood Council Technical Report, no. 10

when used in mass timber applications. Others are created with aesthetics or ease of installation in mind.

5.2.5 FIRE RESISTANCE

Many mass timber products are large, thick, airtight masses of wood. These properties are inherently fire-resistant. This may seem counterintuitive because it is easy to think of wood as a combustible material. However, test results have proven that large wooden components maintain their structural integrity for extended periods, even when exposed to direct flame and intense heat. Fire ratings represent the length of time a given assembly can be exposed to high temperature conditions before losing critical performance characteristics. Design teams will also need to review and address flame and smoke spread classifications (as defined by the IBC) for exposed wood surfaces.

Charring

When exposed to fire, wood chars on its exterior, creating a barrier between the inner portion of the beam/panel and the flame. With continued heat,

the char layer thickens at a very slow, predictable rate, and with each passing moment further insulates the wood at the core. The thickening char layer is removing oxygen from the inner depths of the wood and is, thereby, extinguishing the burning component. This enables the inner, uncharred core to remain structurally unaffected, allowing the component to retain much of its original strength.

The IBC references the NDS for Wood Construction produced by the American Wood Council (AWC) to calculate the fire resistance of mass timber elements (see **Figure 5.14**). This standard establishes a nominal design char rate of 1.5 inches per hour. "Effective" char depth includes a 0.3 inch pyrolysis zone, where the wood is not yet burned but is heated to the point of losing all moisture and is no longer structurally viable. The effective char rate per hour slows the longer wood burns, as the char layer insulates the remaining wood from further damage.

The NDS char rate value is a conservative one. Actual char rates depend on species. Generally, denser (heavier) woods will char at lower rates, while less dense (lighter) species will char faster. The char rates will also depend on species-specific extractives (wood molecules that are non-structural in nature) content, some of which are highly flammable and may accelerate the burning process.

For example, the design team for the Ascent tower in Milwaukee, Wisconsin, opening in 2022, demonstrated that the tall timber structure would have a slower char rating than the prescriptive code value of 1.5 inches per hour. They tested their KLH-supplied panels at the Forest Products Laboratory (FPL) in Madison and found a char rating of 1.29-1.31 inches per hour, saving the project the cost of almost a quarter inch of fiber from every exposed, rated wood component. This finding has excellent implications for design teams pursuing a performance-based permitting process to reduce fiber and costs.

Flame and Smoke Classifications

Interior finish surfaces are classified in the code based on a "flame spread" and "smoke-developed" index with three levels of distinction: Class A is the most resistant; Class C the least. Untreated wood falls into Classes B or C; designations are by species.[17] Flame spread ratings can be improved with treatments and coatings.

Encapsulation

If a design requires fire resistance in addition to the values provided by the wood itself, structural encapsulation is the most straightforward approach, from a code perspective. Under this method, fire safety is attained by encapsulating mass timber elements with an approved and rated assembly. The encapsulation rating is defined as the time that charring of a structural mass timber element is delayed by the "encapsulation membrane," therefore limiting the growth and spread of fire. Gypsum board, gypsum-concrete, and intumescent coatings are among the most popular encapsulation materials.

Coatings

Intumescent coatings and sealants fill gaps and protect the materials underneath them by expanding when exposed to extremely high temperatures. These treatments decrease the immediate flammability of wood, minimizing fuel for an active fire, and slow the spread of a flame. Intumescent coatings can be costly to install, but the thinness and transparency of the coatings solve some dimensional and aesthetic issues, offsetting the cost for some projects.

Exterior Walls

Mass timber components are allowed in exterior wall assemblies, though only up to 40 feet (60 feet if using fire retardant treated wood). The Tall Wood Provisions of the 2021 IBC insinuate the potential to go higher with language allowing timber in exterior wall assemblies for taller buildings, but these provisions are not yet reconciled with limitations in other sections of the code. Testing scheduled for completion in spring 2022[18] will explore the exterior fire spread performance of 3-ply CLT (exposed on the interior side) in non-load-bearing conditions, with the goal of providing data for use in future iterations of the code.

17 American Wood Council, *Design for Code Acceptance 1 (DCA1): Flame Spread Performance of Wood Products Used for Interior Finish* (2019).
18 Timberlab et al., *Wood Innovations Grant: Cross Laminated Timber Exterior Wall Testing to NFPA 285 Test Standard* (USDA Forest Service, 2021).

PMX 15
Source: Sidewalk Labs

CASE STUDY: PMX 15

MANUFACTURED TYPE IV-B BUILDING DESIGNED FOR A HIGH-SEISMIC REGION

LOCATION: SEATTLE

COMPLETION DATE: 2021 (DIGITAL PROTO-MODEL)

ARCHITECT: GENSLER

DESIGN TEAM: ASPECT STRUCTURAL ENGINEERS, INTERFACE ENGINEERING, VORTEX FIRE, AERCOUSTICS ENGINEERING, RDH BUILDING SCIENCE, CADMAKERS, ATELIER TEN

IN 2021, SIDEWALK LABS released PMX 15, a feasibility study for a 15-story timber structure made from a kit of parts specifically designed for off-site manufacturing. As the latest proof of concept in Sidewalk's Proto-Model X series, PMX 15 was developed through 100 percent of the design stage, based on requirements for a building located in Seattle, Washington. PMX 15 focused on three key explorations: maintaining integrity in a high-seismic region; designing to 2021 International Building Code (IBC) criteria; and enabling significant reductions (80 percent) in operational energy use.

Source: Sidewalk Labs

SEISMIC INTEGRITY

To resist lateral movements and forces during seismic events, the PMX 15 team designed a lateral system with steel bracing distributed throughout the building. That's called an "eccentrically braced frame." In the event of an earthquake, the ductile links connecting the steel beams would dissipate energy, directing the movement forces through the link connections in a controlled manner and preventing these forces from reaching and damaging other parts of the building. The eccentrically braced frame enables the building to withstand a 1-in-2,500-years earthquake, as required by seismic code.

2021 IBC CODE

PMX 15 was designed to 2021 IBC IV-B criteria. Most of the beams and columns have some exposure in shared spaces and residential units, but the facades and most of the timber elements within the floor cassettes are encapsulated. For the cassettes, encapsulation made the most sense to address other considerations, such as acoustics and concealment of the in-unit mechanical, electrical, and plumbing systems. All structural elements, whether exposed or encapsulated, achieve the 2-hour fire rating required to meet code.

CARBON IMPACT

PMX 15 achieves an estimated 80 percent reduction in operational energy use relative to ASHRAE industry standards. These gains result largely from the ability to manufacture thermally efficient facades that meet high-performance Passive House building standards and that enable the use of a more efficient radiant heating and cooling system. In addition to operational gains, the embodied carbon impact of PMX 15's structure is roughly 34 percent less than a comparable concrete structure (not counting sequestration). ⬤

5.2.6 STRUCTURAL PERFORMANCE

Foundations

Wooden buildings are much lighter than buildings of a similar size made from steel, concrete, or masonry. Lighter-weight buildings transfer less load to their foundations, leading to smaller, less complex below-grade work, saving on excavation and concrete costs. This is particularly advantageous for building sites with poor soil-bearing pressures, and it also improves the ability to build over contaminated soils with minimum disruption. In one project that required deep foundation piles for an all-concrete building, DCI Engineers was able to realize a 30 percent savings in foundation costs by replacing the top three floors of the building with mass timber.[19]

Using less concrete is desirable for lowering a building's embodied carbon footprint, and it often has significant schedule advantages as well.

Grid Layout/Structural Bay

Mass timber panel dimensions and thicknesses, and, thus, properties of strength and stiffness vary by manufacturer and product. Often, vibration, which in the United States is a subjective value, will govern panel thickness over strength and fire resistance. A design team considering mass timber for floor panels should understand structural bay options and constraints during early building layout decisions.

The manufacturing dimensions of various mass timber panel systems should be considered to optimize use of the material in plan layouts for cost efficiency. It is advisable to bring a procurement or manufacturing partner on to the team as early as possible to gain the benefits of efficient material use. See Section 5.2 for further discussion and Chapter 8 for considerations when advising building owners on contract options.

Seismic Performance

Some of the oldest wooden buildings in the world are in Japan, the most seismically active country on Earth. At over 122 feet tall, the Horyuji Temple, near Osaka, has survived over 46 earthquakes of a magnitude 7.0 or greater on the Richter scale since its construction in 607 AD. Japanese scholars describe the inherent flexibility in these wooden structures as "Snakedance" theory[20], enabling them to dissipate significant seismic energy without damage.

Building codes are the main tool for addressing seismic risks with design requirements, varying by region and depending on the historical frequency and magnitude of earthquake activity. The main seismic criterion in building codes is a specification of the minimum lateral force a building must withstand to assure occupant safety. Building codes include an equation in which cyclic seismic forces are represented by a single static force, called *base shear,* applied to the base of a building. Designers adjust, or design for, variables in the base shear equation to achieve desired building performance. The variables include site seismicity, soil conditions, structural systems and building materials, building height, and building occupancy.

Wood, particularly mass timber, as a structural building material has several characteristics that lead to favorable earthquake performance: ductility, weight, and redundancy.

---

19 1 De Haro, San Francisco, Dean Lewis, DCI Engineers.
20 https://web-japan.org/nipponia/nipponia33/en/topic/

Ductility is the extent to which a material or building can deform without failing. Wood can withstand high-intensity, short-duration loads without failing. Buildings made from wood often use connection systems for joining walls, beams, and columns that further add to a building's ductility.

In high-seismic regions in the United States, building codes limit the use of CLT to resist lateral forces from earthquakes, given the low ductility of the CLT shear wall system (structural R-value of 2). The higher the structural R-value, the lower the lateral force the building is required to be designed to by the building code. Therefore, structural engineers typically design with lateral systems having a higher R-value, such as light-frame timber plywood shear walls (up to R-7).

CLT shear walls and CLT diaphragms now have design requirements defined in the AWC's *Special Design Provisions for Wind and Seismic (SDP-WS)* 2021 edition.[21] This reference guide can be used as a basis for alternative requests to jurisdictions that do not yet recognize IBC 2021. The CLT diaphragm requirements in SDPWS 2021 are engineering-based, with no specific prescribed details. It does include a low-seismic, CLT shear wall option with an structural R-value of 1.5, as well as design details for a platform-framed CLT shear wall system, including specific connectors and aspect ratio limits for individual CLT panels. WoodWorks has a *CLT Diaphragm Technical Guide* that includes working examples using the new CLT diaphragm requirements.

Recent research and testing of CLT shear walls have resulted in proposals to use a structural R-value of 3.0 to 4.0, depending on the aspect ratio of the CLT wall. This, however, still means designing for lateral forces roughly twice that of light-frame plywood shear walls. The R-values of 3.0 and 4.0 for the platform framed CLT shear wall system will be published in ASCE 7, 2022 edition.

Research is ongoing on higher R-value, lower design force, and shear wall systems, including the mass timber rocking wall work led by Shiling Pei of Colorado School of Mines.

Weight: Lighter building weight is an advantage in a seismic event because the inertial force exerted on a building is proportional to weight, with higher inertial forces exerted on heavier buildings. Lateral systems for timber buildings are required to resist less force than heavier buildings, and as a result can be smaller and less expensive.

Redundancy: Many fasteners and connectors are used in wooden buildings to join walls, roofs, floors, beams, and columns. Each of these connections is a load path through which seismic forces can travel. The numerous connections inherent in a component-based construction approach mitigate the chance for complete structural failure if some connections fail.

Wind Loading

In regions with low seismic concerns, or in very tall buildings, wind loads may govern lateral design. Many of the timber advantages discussed in the seismic performance section can be applied to wind loading design. However, lighter-weight buildings will require adapted shapes and/or more lateral strengthening forces than heavier buildings to deflect or resist wind.

---

21   https://awc.org/codes-standards/publications/sdpws-2021

HYBRID SYSTEMS

Most timber structures use steel-reinforced concrete for foundations and steel components for connections. A project that uses a full-building hybrid approach, however, efficiently combines multiple primary structural materials. Factors such as building height, grid layout, and seismic region may lead a design team toward a hybrid building approach. Although wood is very strong by weight in both tension and compression, selectively incorporating concrete or steel, or a combination of both, can mitigate vibration, increase span capacity, reduce structural member dimensions, or increase lateral capacity. Whole buildings are often hybrid designs, but component-based approaches, such as hybrid slabs and lateral systems, are also being developed in research and in practice (see **Figures 5.15** and **5.16**).

Hybrid Slabs

Some building programs require spans that are difficult to accomplish with mass timber panels alone.

TOP — FIGURE 5.15: CONCRETE CORES AND PRECAST CONCRETE FRAME WITH TIMBER SLAB AND BEAMS

*Adidas North American Headquarters, Portland, OR
Source: Lever Architecture*

BOTTOM — FIGURE 5.16: HYBRID CLT AND STEEL STRUCTURE

Microsoft Campus, Mountain View, CA; Source: Holmes Structures; Photo Credit: Blake Marvin Photography

An efficient classroom building on a 30-foot grid, for example, might at first seem to call for solid timber floors with a thick section that's cost-prohibitive. For such projects, designers may instead consider adding beams, tension cords, or composite slabs, or they could rethink standard grid approaches developed with other construction materials. Established options for hybrid slabs include:

TOP — FIGURE 5.17: PEAVY HALL: EXAMPLE OF COMPOSITE CONCRETE-TIMBER SLABS

Peavy Hall, Oregon State University
Source: Evan Schmidt

BOTTOM — FIGURE 5.18: UMASS AMHERST: EXAMPLE OF COMPOSITE CONCRETE-TIMBER SLABS

John W. Oliver Design Building at UMass Amherst
Source: Alex Schreyer/UMASS

- Composite concrete-timber slabs are composed of concrete and timber connected via steel components to create composite structural action; they take advantage of the properties of both materials simultaneously. A concrete diaphragm is poured over a timber slab and connected with reinforcing steel to tie the two materials together. Thickened concrete sections may act as beams. Reinforcing steel can take many inventive shapes, such as fasteners driven into the timber at an angle before the concrete is poured (see **Figure 5.17**), perforated steel flanges added during the timber manufacturing or glued in on-site (see **Figure 5.18**), or two-way rebar. In Europe, special types of removable anchoring systems are being developed to facilitate the deconstruction of concrete-timber slabs at

TOP LEFT — FIGURE 5.19: POST TENSIONED TIMBER BEAM

Source: 120 Clay Creative, Ankrom Moisan
Photo Credit: Ethan Martin

TOP RIGHT — FIGURE 5.20: POST TENSIONED CLT PANEL

Chibougamau terminal, Nordic Structures and EVOQ
Architecture; Photo Credit: EVOQ/Artcad

BOTTOM — FIGURE 5.21: TIMBER-TIMBER
COMPOSITE FLOOR PANEL

Catalyst, Katerra; Photo Credit: Andrew Giammarco

FIGURE 5.22: LIGHT FRAME AND MASS TIMBER HYBRID

The Canyons, Portland, OR
Source: Kaiser+Path; Photo Credit: Marcus Kauffman, Oregon Department of Forestry

the building's end of life. Permanently integrated concrete-timber slabs may be very difficult and expensive to handle during deconstruction, weighing on the cradle-to-grave carbon balance of buildings that use them.

- Post-tensioned timber: Adding steel tension cords to timber beams can reduce overall beam depth and increase structural transparency (see **Figure 5.19** and **5.20**).

- Timber-timber composite floor panel: Timber slabs with thickened timber sections can increase span capacity. Catalyst, an office building project in Spokane, Washington, conceived and developed a timber-timber composite floor panel to achieve a 30-foot span with CLT floors and shallow CLT beams integrated during panel fabrication (see **Figure 5.21**). Other manufacturers have produced mass timber hollow core

panels that combine thinner (3-ply) CLT panels for top and bottom layers, connected with internal glulam ribs. The hollow spaces are filled with insulation materials. Mass timber ribbed panel assemblies are another relatively new mass timber product combining CLT decks with integrated glulam ribs connected by screws, glue, or a combination of both to the bottom.

Hybrid Lateral Systems

Because of the stiffness of mass timber panels (see Seismic Performance: Ductility section above), using hybrid approaches for lateral systems is often cost-effective. Common strategies include:

- For mid-rise structures, light-framed wood shear walls are a straightforward and cost-effective approach (as in **Figure 5.22**).

FIGURE 5.23: CLT POST AND BEAM STRUCTURE WITH BUCKLING RESTRAINED BRACED FRAME CORE

Carbon 12, Portland, OR; Source: Kaiser + Path

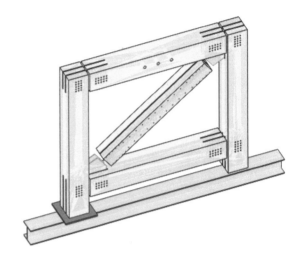

FIGURE 5.24: TIMBER BRB FRAME TEST SCHEMATIC

Source: Timberlab

nents as a kit of parts for rapid on-site assembly in any weather. An all-timber BRB frame lateral system that works in high seismic zones is under development at the University of Utah. Final testing is scheduled for spring 2022.[22]

- For taller buildings, concrete cores can be advantageous from a permitting and constructability perspective. Concrete cure times should be considered and construction sequencing optimized so building the cores does not offset the time-saving advantages of timber framing.

- Buckling Restrained Braced (BRB) frame cores and walls, which can be prefabricated with steel (as in **Figure 5.23**) or glulam (as in **Figure 5.24**) cross bracing, have time-saving advantages over concrete in construction. BRB frames can be designed with bolted rather than welded connections, working with the mass timber compo-

- Post-tensioned CLT shear walls combine strong, rigid wood panels with steel tendons and fuses for added ductility and seismic force dissipation (see also Chapter 8's section on resiliency). The technology was developed in New Zealand and has been in use there for nearly a decade. Peavy Hall, at Oregon State University, is the first installation in North America (see **Figure 5.25**).

5.2.7 ACOUSTIC PROPERTIES

Mass timber has advantages as an acoustic solution. The massive arrangement of wood helps mitigate transfer of low-frequency sound vibrations. Combining mass timber with other build-

22 University of Utah, et al., *Wood Innovations Grant, BRB Braced Frames for Seismically Resilient Mass Timber Buildings*, (USDA Forest Service, 2021).

FIGURE 5.25: POST-TENSIONED CLT 'ROCKING' SHEAR WALL INSTALLATION

Peavy Hall, Oregon State University; Photo Credit: Hannah O'Leary

ing materials can create relatively thin assemblies with high Sound Transmission Class (STC) and Impact Insulation Class (IIC) values.

Some standard assemblies for acoustical performance in mass timber buildings have been developed, as well as an array of proprietary solutions. WoodWorks has an online inventory of hundreds of mass timber assemblies that have been acoustically tested.[23] Additionally, some guidelines have been developed for floor assemblies, including a mass timber floor with a raised access floor, a mass timber "dry" build-up, and numerous as-

semblies specific to the 2021 IBC tall mass timber construction types.

A 2019 research project[24] at TDI showed promising outcomes for five common floor assemblies, each with a CLT and MPP iteration (see **Figure 5.26**). STC and IIC values were above 50 for all floor assemblies with acoustic underlayment and floating floors, except for IIC values on assembly F05, a dry assembly with tongue-and-groove engineered pine flooring. STC and IIC values for bare timber assemblies and bare timber-compos-

23 https://www.woodworks.org/wp-content/uploads/Acoustically-Tested-Mass-Timber-Assemblies-WoodWorks.pdf
24 Kevin Van Den Wymelenberg, "Acoustic Testing of Typical Multi-Family Residential CLT and MPP Dry and Concrete-Composite Wall and Floor Assemblies."

CLT + MPP FLOOR TESTING RESULTS

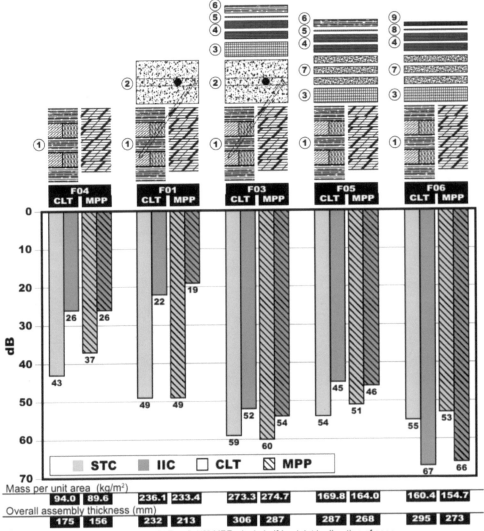

(1) Mass timber panel, 6-7/8" 5-lam CLT or 6-1/8" MPP, single half-lap joint in direction of span

(2) 2-1/4" concrete slab @ 145 pcf density, #3 rebar, 6" o.c. in span direction, 12" o.c. perpendicular to span, 8mm x 220mm shear fastener, 12" o.c. field spacing @ 45°

(3) 1" acoustic underlayment, install in opposing direction to flooring

(4) 5/8" OSB, 2 layers, glued in direction of span, stagger seams, adhere with construction adhesive

(5) 1/8" acoustic underlayment

(6) 6-1/2" x 1/2" random length engineered pine floating floor, T&G, sanded, oiled

(7) 1/2" cement board, 3 layers, stagger seams, adhere with construction adhesive

(8) 3/8" 8lb carpet pad

(9) 1/2" pile nylon carpet, 97.5 oz/sq.yd face weight

UNIVERSITY OF OREGON | Energy Studies in Buildings Laboratory

FIGURE 5.26: CLT + MPP FLOOR ASSEMBLY ACOUSTIC TESTING

Source: University of Oregon, Acoustic Lab Testing (ASTM E492-2016, ASTM E90-2016) of CLT and MPP Wall and Floor Assemblies for Multi-Family Residential

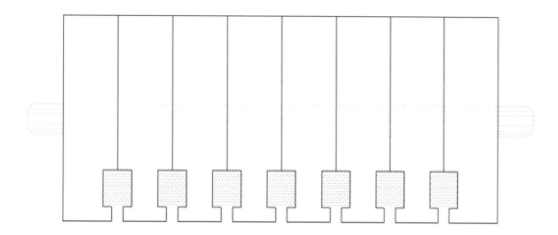

FIGURE 5.27: SIDE VIEW OF ACOUSTICALLY DESIGNED DLT PANEL

Source: StructureCraft

ite assemblies fell below 50, but STC values were 49 for bare concrete-timber composite floors.

As with other code-required assemblies, permitting authorities may allow a performance-based approach for acoustic ratings. An acoustic engineer can review floor and wall assemblies, make performance recommendations, and provide project specific STC and IIC values.

Some mass timber panels are specially designed for acoustic performance. For example, StructureCraft produces a sound-dampening DLT panel with insulation-filled grooves engineered to absorb sound waves (see **Figure 5.27**). In Europe, many panelized mass timber projects use special durable polymer dampening seals in pedestal-type, floor-to-wall connection to further reduce vibration and sound transmission in buildings.

5.2.8 THERMAL PERFORMANCE

The thermal performance of a building directly influences not only its energy efficiency but also the occupants' comfort and the lifespan of some building components. Mass timber is an excellent material selection for thermal performance. Wood is a good insulator and is universally appealing, with exposed wood surfaces giving occupants a "warm" feeling (see Chapter 7 for more detail on occupant comfort).

The thermal performance of a building is dependent on many factors, including climate, building shape, building orientation, architecture, and building and insulating materials. The R-values and k-values of various building materials help determine the overall thermal performance of a structure. The k-value, known as "thermal conductivity," is a measure of the rate of heat transfer through a material. The unit of measure for this rate is watts per meter kelvin; the measure is independent of the material's thickness. Materials with high thermal conductivity transfer heat more quickly, and thus are generally not useful insulators. Materials with low thermal conductivity transfer heat more slowly and are more likely to be found in insulating applications.

MATERIAL	THERMAL CONDUCTIVITY K-VALUE (W/(M K))
Sheep wool	0.04
Insulation, average quality	0.04
Sawdust	0.08
Douglas fir	0.12
Hemlock	0.12
Plywood	0.13
Southern Yellow Pine	0.15
Gypsum board	0.17
Plaster and wood lath	0.28
Concrete, medium	0.4 – 0.7
Concrete, dense	1.0 – 1.8
Steel, 1% carbon	43.00

TABLE 5.1: THERMAL CONDUCTIVITY OF BUILDING MATERIALS

Source: Engineering Toolbox, Thermal Conductivity of Common Materials and Gases (2003)

material. **Table 5.1** shows some common building materials (and other materials for comparison) and their thermal conductivity values.

Solid wood has relatively low thermal conductivity and can, therefore, be used as an insulator. The thermal conductivity of solid wood is up to 15 times lower than concrete, and over 350 times lower than steel. Mass timber buildings can be designed and built with superior thermal performance, leading to reduced energy requirements over the life of the building. This provides cost savings for building owners and occupants, and it reduces the operational carbon footprint.

Air infiltration rates of exterior envelopes also contribute significantly to the energy performance of a building. CLT has an exceptionally low air infiltration rate, making it a good choice for the high-performing exterior walls required for very low-energy building design.

5.2.9 MOISTURE

A mass timber designer will need to consider concerns like those associated with light-frame construction and finish wood products, but there are also a few key differences, outlined in this section. Understanding wood's behavior as an organic material is foundational to establishing best practices.

The thermal R-value, known as "thermal resistance" (not to be confused with the structural R-value discussed in an earlier section), can be measured for an individual layer of material. It quantifies the effectiveness of that layer as an insulator, given its thickness. Thermal R-value is calculated by taking the thickness of a layer and dividing it by the thermal conductivity of the

Wood has a cellular structure ideal for holding and distributing moisture within a live tree. Once harvested, wood fibers continue to be hygroscopic, readily expanding and contracting as environmental moisture content increases or decreases. Controlling the moisture exposure of wood building products is important along the entire supply chain, from lumber processing to fabrication, delivery, construction, and occupancy. Maintaining a relatively stable moisture content at each stage avoids the performance and aesthetic concerns that arise from dimensional changes, cracking or checking, staining, and decay. Factors most commonly contributing to these issues are exposure to weather before or after occupancy, trapped (unventilated) moisture, and roof or plumbing leaks.

The moisture content of logs at harvest may exceed 100 percent (i.e., there may be more water than dry woody substance) by "oven dry base," the metric used by the lumber industry. Of the total weight of the water in a log, about 60 percent is "bound" within the anatomical structure of individual cells. The balance is "free" water in cavities within the wood cells. For the types of lumber used to make mass timber, industry expectations are that the lumber will be dried to 12 percent moisture (+ or − 3 percent). Drying lumber to this level helps assure dimensional stability during mass timber manufacturing and use, and it prevents decay. The ideal moisture content for fungal growth ranges between 26 percent and 60 percent. Factors contributing to the variances include wood species, fungus species, temperature, and time (rate of dry out).

In wet climates, wood absorbs moisture during the construction phase, and a building must go through a "dry-out" phase before wood is enclosed—or risk compromise. A building with properly ventilated and dried wood will stabilize during the first two or three years of occupancy to match the ambient moisture content, typically 6 percent to 8 percent for wood in interior use applications in the Pacific Northwest. The greater the moisture content differential within a wood member, or between the installed wood and the future occupied building, the greater the impact of shrinkage and checking.

Ongoing research in academia and industry will continue to inform the best practices for protection and detailing. Although industry standards are nascent for many of the issues specific to mass timber and moisture mitigation, resources for designers are developing. In early 2020, RDH Building Science Inc. published advice to designers on some aspects of detailing mass timber buildings to protect and recover from moisture exposure.[25]

A "Water in Mass Timber"[26] project is ongoing at TDI via grants from the United States Department of Agriculture (USDA) and the Agricultural Research Service. One aspect of this project is exploring the effects of a variety of moisture exposures (ambient exposure through sustained flooding) on the performance of timber connections and providing benchmark data for engineering models. In early 2020, hundreds of connection samples were prepared and inoculated with 2 decay fungi, and the first set of specimens was harvested in early February 2021. Some testing of water-exposed connections has been completed and results will be available soon.

Moisture Management and Monitoring

Specifications should include expectations about weather protection for stored and in situ materials during construction. A moisture management plan should be in place before construction starts, and a clear strategy should be proposed before building costs are finalized (see also Chapter 6). Monitoring moisture before and during dry out

25 RDH Building Sciences, *Mass Timber Building Enclosure Best Practice Design Guide* (2020).
26 PI Arijit Sinha, "Water in Mass Timber," Oregon State University (OSU).

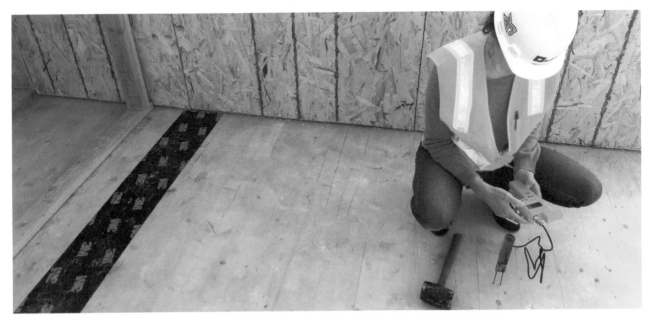

FIGURE 5.28: MOISTURE MONITORING CLT FLOORS WITH A HAMMER-IN PROBE

Source: Kaiser + Path; Photo Credit: Kevin Lee

with an instrument designed to measure wood moisture content (see **Figure 5.28**) will validate if panels are ready to be enclosed or encapsulated with other materials.

Massive panels dry at different rates than stick framing (see Chapter 6 on weather and weather protection for more information), and the dry-out period should also be considered in terms of both schedule and technique. Allowing wood to slowly reach moisture equilibrium mitigates potential shrinkage and checking issues, of special concern where structural wood doubles as a finished surface.

Mitigation

The most effective and low-cost way for a designer to protect a wood building from moisture is through architectural detailing. Treatments or coating products add to the cost and environmental footprint of a project, so they should be used sparingly, but they may be warranted to protect against various exposure conditions.

- Proper architectural detailing, with little to no additional cost, incorporates expansion joints to allow for shrinkage, considers protection from direct moisture contact, and allows wood in place to breathe (release moisture). These details should also protect wood from exposure and contact with materials like concrete that can transfer moisture. Designers should consider that moisture is absorbed and expelled most rapidly through the wood's end-grain, and that most shrinkage happens tangentially or radially (see **Figure 5.1**).

- Wood coatings can add protection against moisture and UV to the completed building or during construction exposure—or both. Mass timber manufacturers often have standard temporary coatings to protect wood during transport, storage, and installation. These products should be included in the specifications for clarity and for coordination with other specified coatings. Coatings that are applied to encapsulate dry wood and keep it from gaining moisture from the external environment work if the coating is not compromised. Any intended cuts or scratches in a coated element create breaches in the encapsulating coat that allow moisture in, while the remaining intact coat prevents drying.

- Treated wood is common for exterior wood structures such as bridges, decks, railroad ties, and telephone poles. Not all treatments are appropriate for occupied structures, as many formulas come with human health risks. Treatments tend to come at a higher cost than coatings, but they are highly effective. Chemical changes at the cellular level may alter the composition of the wood, and, in some cases, can also diminish strength properties. The mass timber market has few options for treated wood, owing, in part, to the large dimensions of mass timber components, but several testing efforts are in progress to analyze the structural performance of treated mass timber and its interactions with adhesives. Treated mass timber panels could also have insect repellent capabilities, expanding geographic acceptance into regions with termites. The large size of the panels makes post-treatment impractical.

Dimensional Stability

Engineered wood elements like CLT are less susceptible to in-plane dimensional changes because of moisture and temperature swings than lumber or sawn timber because adhesives and multiple fiber directions hold their overall dimensions (excepting panel thickness) stable. CLT and MPP panels, therefore, have an advantage over NLT or DLT if a building is constructed during wet weather. Potential dimensional changes during construction in wet climates should be factored in when detailing these systems.

The smooth, precise look of a freshly pressed CLT panel is more likely to be preserved if moisture content is stable from manufacture through installation. A CLT panel is manufactured with little to no gap between each board in a lamination. If a CLT panel becomes saturated, the added moisture can cause each laminated board to swell and push against the others, while the overall panel width and length dimensions remain stable. The more significant the drop in the moisture content of a panel, the larger the gaps between each board—or cracks in the case of edge-glued boards—will be. Some European-sourced panels have edge-glued boards to eliminate shrinkage gaps at each board seam. Because CLT adhesives are stronger than wood fiber bonds, shrinkage cracks then occur within boards, rather than between them, as a panel takes on and releases moisture.

Building Shrinkage

Cut wood contracts and expands differently depending on its relationship to the growth rings and the direction the fiber runs. Radial and tangential dimensions change more significantly than in the direction of the grain. In light-wood framing, shrinkage is calculated mostly within the top and sill plates, while vertical wall studs contribute very little to potential building shrinkage.

Mass timber elements will contribute to prevention of shrinkage, depending on the detailing and the products used. If used for floors, for exam-

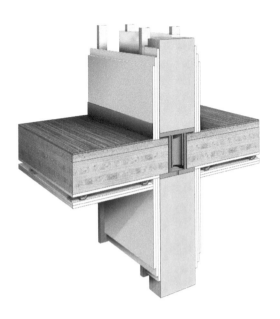

FIGURE 5.29: END-GRAIN TO END-GRAIN COLUMN CONNECTIONS MINIMIZE SHRINKAGE

Brock Commons, University of British Columbia
Source: Acton Ostry Architects

5.2.10 BALLISTIC/BLAST PERFORMANCE

The US military is interested in using mass timber in construction projects, with one estimate finding that military construction using CLT instead of concrete and steel could create a market of $1.9 billion annually for buildings, housing, and facilities requiring low levels of blast resistance.[27] When designing military buildings, architects are often required to integrate blast- and projectile-resistant materials.

Initial blast resistance tests conducted at Tyndall Air Force Base in Florida validated acceptable levels of blast resistance for structures built with NLT and CLT. All structures remained intact and matched modeling predictions for acceptable levels of damage. Additional testing is underway.

In addition, efforts are underway to understand how mass timber structures perform when struck by projectiles. Georgia Institute of Technology (Georgia Tech) has completed studies in which CLT panels made of spruce, pine, fir (SPF) and Southern Yellow Pine (SYP) were subjected to ballistic testing. The results showed that both types of conventional CLT materials' inherent penetration resistance is significantly greater than that of the dimension lumber and plywood now used for temporary military structures. Additionally, the testing showed that US military guidelines (UFC 4-023-07) for determining required wood thickness based on ballistic threat underestimated the performance of CLT. The tests resulted in new equations for predicting the required thickness of CLT for ballistic protection.[28]

ple, CLT will contribute to shrinkage in a platform-framed building using CLT as floors, while this effect could be avoided with a balloon-frame approach. Because shrinkage in the direction of the grain is almost negligible, shrinkage can be largely avoided with details that utilize end-grain to end-grain connections. For example, both the 18-story Brock Commons at the University of British Columbia (UBC), shown in **Figure 5.29**, and the 8-story Carbon12 in Portland were designed with stacked glulam columns with steel connections in between. This has more impact in taller buildings, where the accumulation of floor-to-floor shrinkage becomes a greater concern because of a greater number of floors.

27 Woodworks, "Cross Laminated Timber Blasts its Way into Government Construction."
28 Kathryn P. Sanborn. Ph.D., "Exploring Cross-Laminated Timber Use for Temporary Military Structures" (thesis, Georgia Tech University).

Source: Heatherwick

CASE STUDY: HESS TIMBER/GOOGLE

HESS SUPPORTS FACADE SYSTEM FOR GOOGLE'S 'LANDSCRAPER'

GOOGLE'S NEW HEADQUARTERS FOR around 7,000 employees is being built behind what will be the world's largest wooden facade (23,300 square meters in total area). Together with German facade specialist Josef Gartner GmbH, HESS TIMBER GmbH (HASSLACHER group) is developing and supplying the facade system for Google's headquarters near King's Cross Station in London. When the facility is completed, Google employees will have access to a spacious park with trees, green spaces, and a running track on the roof of the so-called "Landscraper." The building's design was developed by internationally renowned architects Bjarke Ingels of BIG and Thomas Heatherwick of Heatherwick Studio.

RENEWABLE BUILDING MATERIALS

The building materials for this project had to be sourced from within Europe. Because of the special requirements for an external, exposed facade element, a hybrid setup consisting of European larch (*Larix decidua*) and acetylated radiata pine (*Pinus radiata*) was chosen. In this case, however, the radiata pine came from Spain, not from Chile or New Zealand, as is usually the case. Initial materials tests showed that the raw material from Spain differed from South American or New Zealand radiata pine, especially in mechanical properties and gluing behavior.

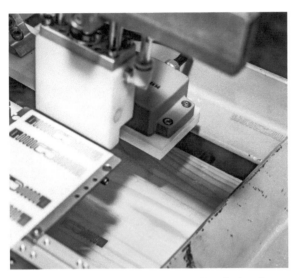

Source: Hasslacher Group

Source: Hasslacher Group

MATERIALS TESTS

European materials institutes conducted extensive tests to develop and execute the hybrid timber solution for the mullions, or window dividers. In addition to acetylated radiata pine, around 1,750 cubic meters of glulam made of European larch (*Larix decidua*) was used for the facade elements; the moisture-induced deformation behavior also had to be scientifically investigated, as no literature was available.

LARGE-SCALE MOCK-UPS

Over several months, various mock-ups on a scale of 1:1 were built on Gartner's test site in Gundelfingen, Germany. The insights gained from these "stress tests" were crucial to the final production of the mullions.

PRODUCTION AND RFID TECHNOLOGY

The production of the glulam and Cross-Laminated Timber (CLT) elements is well underway, and some have been shipped to the UK and installed on-site. Each mullion is equipped with nine radio frequency identification (RFID) moisture sensors that detect when surface water accumulates. Once installed, the RFID sensors can be maintained by being inductively scanned with a special readout device.

JUST-IN-TIME DELIVERY

A sophisticated logistics concept had to be established to cope with the sheer quantity of elements, as well as external factors such as the COVID-19 pandemic, travel restrictions, an ever-more-complicated logistics chain, and timber facade elements that must arrive on-site in the correct sequence and time.

OUTLOOK

In this project, experiences in the fields of research and development were deepened, including those of renowned research institutions. Thus, new insights into the use of modified woods for outdoor applications were developed. The exposure and visibility of this unusual wooden facade can contribute to a broader acceptance of the most natural, renewable building material—wood. ⬤

5.3 PROJECT MANAGEMENT AND COORDINATION

In these early stages of the introduction of mass timber to North America, design teams need to be well educated about how to best integrate the many benefits of these products into their projects. Development teams must include architects and engineers who know well the advantages and disadvantages of these products. CLT is not simply a replacement for concrete. They both have very different characteristics and design considerations.

5.3.1 PLANNING AHEAD

Design-phase-forward planning can result in significant improvements to construction schedules, but it requires more coordination earlier in the design process. Project managers should account for this when advising owners, determining fees, scheduling staffing, and choosing consultants and software tools. More coordination time before construction starts can reduce costly field labor and project overhead costs, and it can deliver a superior product.

Design Partners

Early Mechanical, Electrical, and Plumbing (MEP) coordination, for example, can have positive aesthetic, cost, and maintenance implications in the final building. Many MEP consultants provide a diagrammatic design, intending for the final layouts to be largely field coordinated. In a mass timber building, however, the structure is often substantially exposed. Thoughtfully exposing utilities where necessary or desired requires working with consultants early on to consolidate utilities in carefully planned zones, and to arrange for higher quality materials in exposed areas. Penetration locations can be determined before timber compo-

nents are fabricated, reducing on-site trade conflicts and speeding installation. Planning for more off-site fabricated components can improve scheduling and craftsmanship, while reducing risks.

Additional benefits to a building owner go beyond aesthetic and construction advantages. In the completed building, as-built reference documents will be more accurate, requiring fewer modifications from the original design documents. Building operations and management teams working with logical, accurate reference materials also will be more efficient and successful.

Procurement and Construction Partners

One of the unique opportunities inherent in designing with mass timber is how the new technology makes clear the stark advantages of an integrated design-and-build team. To produce an efficient and cost-effective mass timber design, the design team ideally works with a procurement team early in the design process able to track and advise on market and supply trends as the building design evolves. A building owner should be advised to use collaborative contract models that support effective prebid coordination (see also section 8.2 in Chapter 8). Working with design-build trade partners can also provide valuable continuity from early design through close-out.

Site coordination concepts and installation approaches can impact estimated costs significantly. A general contractor who can calculate the cost savings achieved by a modular mass timber approach can advise on overall construction schedule reductions when compared with other construction techniques. Choosing a construction partner who is familiar with the unique time and cost savings mass timber can offer is key to real-

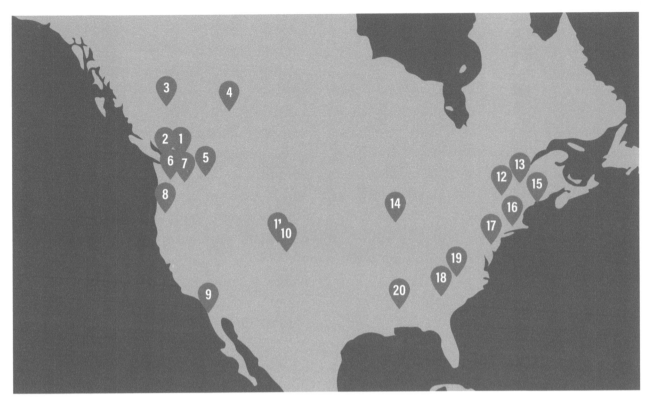

FIGURE 5.30: INDUSTRY-COLLABORATIVE WOOD RESEARCH INSTITUTIONS IN NORTH AMERICA

izing those savings in early cost models or bids. Structural mass timber panel manufacturing companies often offer architectural and engineering design, modeling, project management and construction as parts of an integrated package, resulting in a building shell. Assistance in all these functions may also be offered to clients selecting external designers and/or contractors.

Research Partners

For novel and performance-based design approaches, design teams can utilize testing and research resources available through collaborative research institutions throughout North America. **Figure 5.30** and the list below identifies nonprofit, building-industry supportive institutions with physical laboratory facilities and expertise in mass timber specific focus areas.

Northwest

1. FPInnovations (Vancouver, BC)
2. University of British Columbia Timber Engineering and Applied Mechanics Laboratory (Vancouver, BC)
3. University of Northern British Columbia The Wood Innovation Research Lab (Prince George, BC)
4. University of Alberta Advanced Research in Timber System (Edmonton, AB)
5. Washington State University Wood Materials & Engineering Laboratory (Spokane, WA)
6. University of Washington Construction Materials Lab (Seattle, WA)
7. APA Research Center (Tacoma, WA)
8. Tallwood Design Institute, Oregon State University (Corvallis, OR) and University of Oregon (Eugene, OR)

Southwest

9. NEHRI Shake Table (San Diego, CA)
10. Colorado School of Mines (Golden, CO)
11. Colorado State University (Fort Collins, CO)

Northeast

12. FPInnovations (Pointe-Claire, QC)
13. Université Laval CRMR Lab (Québec, QC)
14. Forest Products Laboratory: USDA Forest Service (Madison, WI)
15. University of Maine Advanced Structures & Composites Center (Orono, ME)
16. UMass Amherst Wood Mechanics Lab (Amherst, MA)
17. Leheigh University (has done some testing)

Southeast

18. Clemson Wood Utilization & Design Institute (Clemson, SC)
19. Virginia Tech Sustainable Biomaterials Lab (Blacksburg, VA)
20. Mississippi State University Department of Sustainable Bioproducts (MS)

5.3.2 BUILDING INTEGRATED MODELING (BIM)

Building Information Modeling (BIM) is a process of creating virtual models built in 3 dimensions, including detailed or approximated components of all the elements that will make up a building. BIM are used for coordination and collaboration across architecture, engineering, manufacturing, and construction fields. In the last decade or so, BIM programs have become standard tools for design documentation in most design disciplines, and they have revolutionized construction coordination and "clash-detection," as well. These developments are

FIGURE 5.31: STRUCTURE DETAIL AT PREFABRICATED OFFICE BUILDING

IZM Building, CREE and Hermann Kaufmann
Source: Emily Dawson

auspiciously synchronized with the development of modular timber construction techniques. Design and construction models can often be adapted into shop drawings, facilitating communication around complex 3D material intersections. BIM models can be highly detailed, so it is possible to have the quantities and dimensions of any building component, from conduits to fasteners to mass timber panels, predetermined well before they arrive on site.

5.3.3 PRECISION AND PREFABRICATION

The precision and design control of prefabricated building components (see **Figure 5.31**) appeals to designers around the world. Prefabrication has many benefits for the construction schedule, as discussed in more detail in Chapters 6 and 8. Because it is inherently prefabricated, designing with mass timber may lead to further discussions of off-site fabrication, allowing it to grow from a focus on structure into more complex systems components, full wall assemblies, or even volumetric modular spaces. A build partner familiar with these techniques is crucial to realizing the potential of more complex prefabricated components. A project's location and the availability of prefabrication facilities will also play a role in cost and viability.

Implications for the design team include, as previously discussed, planning for more up-front coordination. The extent of prefabricated components—and how they are sourced, manufactured, and procured—will dictate the amount of extra coordination required.

5.4 BUILDING CODES

Historically, common wood structural building materials and methods have been included in building codes across North America using Type IV construction. Type IV allows for the use of heavy, solid sawn timbers (6 inches and larger in vertical framing components and 8 inches and larger in horizontal components), as well as commonly available wood composites, such as glulam beams. Other construction types (I, II, III, and V) allow for the use of wood elements in places, though additional protection may be required to increase fire resistance.

When a building material or construction method is not included in applicable building codes, any building project team desiring to use that material or method must have the building permitted using an "alternate means" approach, to demonstrate to the permitting body that the materials and methods are at least equivalent to adopted codes for the specified use. This process can be costly, time-consuming, and difficult, and it does not have a guaranteed outcome. Including newly developed mass timber products and methods in building codes removes significant barriers to that product or technology's adoption in the marketplace. Although organizations in the US and Canada develop building codes at the national level—the International Code Council (ICC) and the Canadian Commission of Building and Fire Codes (CCBFC)—it is up to state/provincial and local authorities to adopt these codes, creating a patchwork effect.

In recent years, several building code changes specific to the use of wood structural components have been made at the national, state or province, and local levels.

5.4.1 2015 NATIONAL BUILDING CODE OF CANADA

The 2015 National Building Code (NBC) allows the use of wood as the structural frame in buildings as tall as 6 stories for residential, office, and mixed-use occupancies. The previous version of the code allowed wood only in residential buildings, and they were limited to 4 stories. This update also recognizes mass timber for use in podiums, considered noncombustible (NC). Two construction types are recognized in this version of the code: 1) combustible (includes heavy timber, but recognized as having NC properties), and 2) NC. Updates to the NBC, which is developed by the CCBFC, come out every 5 years and are adopted on a province-by-province basis. Most regions in Canada have adopted the 2015 code.

5.4.2 2020 NATIONAL BUILDING CODE OF CANADA

The 2020 update of the NBC adds a new construction type: Encapsulated Mass Timber Construction, commonly referred to as the "EMTC provisions." The new code increases the maximum allowable height of mass timber structures from 6 to 12 stories. Requirements include encapsulation of structural timber with noncombustible materials, and limited permissions for exposed struc-

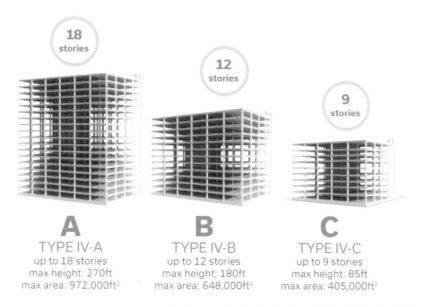

A

TYPE IV-A
up to 18 stories
max height: 270ft
max area: 972,000ft²

B

TYPE IV-B
up to 12 stories
max height: 180ft
max area: 648,000ft²

C

TYPE IV-C
up to 9 stories
max height: 85ft
max area: 405,000ft²

FIGURE 5.32: TALL WOOD CONSTRUCTION TYPES ADDED TO THE 2021 IBC

Think Wood Research Brief Mass Timber 2021 Code

tures. The 2020 NBC is expected to be approved for adoption by the end of 2021.

5.4.3   2015 INTERNATIONAL BUILDING CODES

In early 2015, the ICC adopted new codes allowing the use of CLT in buildings up to 6 stories for offices and 5 for residential. However, CLT use in taller buildings was not addressed in this code update. Because CLT is viewed as having the most competitive advantages (in terms of cost and appropriateness of application) in buildings that are 6 to 16 stories tall, the 2015 IBC adoption was considered only a partial improvement. IBC updates are adopted on timelines determined on a state-by-state basis.

5.4.4   2021 INTERNATIONAL BUILDING CODES

The 2021 edition of the IBC includes major changes to Construction Type IV specific to mass timber. They include provisions for the use of mass

timber as a primary structural material in buildings up to 18 stories in height. These changes are often referred to as the "Tall Wood Provisions."

Construction Type IV was revised to IV-HT, and now also includes three additional types, distinguished by fire resistance, height, and area restrictions (see **Figure 5.32**).

Type IV-HT: Maximum 6 stories, 85 feet in height, and 108,000 square feet in area. Concealed spaces are now allowed with exceptions for sprinklers, filled cavities, and protection with NC construction, like gypsum.

Type IV-C: Maximum 9 stories, 85 feet in height, and 405,000 square feet in area, and all mass timber designed for a 2-hour fire resistance may be exposed. Concealed spaces are allowed if protected with NC.

Type IV-B: Maximum 12 stories, 180 feet in height, and 648,000 square feet of area. Exposed mass timber walls and ceilings are allowed with limitations, and concealed spaces are allowed if protected with NC.

Type IV-A: Maximum 18 stories, 270 feet in height, and 972,000 square feet in area. NC fire protection is required on all mass timber elements, and concealed spaces are allowed if protected with NC.

This groundbreaking advancement of the code is a huge step for the uptake of mass timber in the US, though the encapsulation requirements have been questioned as conservative. Because the requirements add costs and diminish many of the benefits of a mass timber building to occupants and owners, they create cost-effectiveness challenges for taller wood structures. This is seen as an urgent area of research for the industry. Through wood innovation grants, testing to reduce the encapsulation requirements of the new code provisions is ongoing.

5.4.5 TALL WOOD CODE ADOPTION

More states in the US have been adopting the Tall Wood Provisions since the 2021 ICC was released. Oregon and Washington have been leaders in the adoption of mass timber construction, proactively adopting the Tall Wood CLT Provisions in 2018. Maine incorporated them into the state building code in January 2021, followed by California and Georgia in July. Utah fully adopted the 2021 IBC in March of 2020, and Virginia amended the 2018 code by adding a reference to the 2021 IBC to include the new types of tall mass timber construction. Idaho is the most recent state to approve the provisions, for use starting in 2022.

Proactive city jurisdictions have been adopting the provisions ahead of state approvals. In December 2019, Denver, Colorado, approved the new provisions for immediate use, and Austin, Texas, adopted them in September 2021. In October 2021, the New York City Council approved the use of mass timber for construction of buildings of up to 85 feet tall.

The City of Vancouver, British Columbia, which recognizes its own code authority autonomously from the province, has adopted the Tall Wood aspects of the 2021 NBC code. British Columbia and Alberta have allowed jurisdictions to apply for early adoption, and dozens have. Ontario has been supportive of alternative equivalent solutions for mass timber projects. Though it has not been considering early adoption, there are several projects over 6 stories planned for construction in the coming year, including an 11-story project in Toronto.

5.4.6 FUTURE CODE UPDATES

Following a successful fire compartment testing study performed in Sweden in 2020[29], which included technical advisory partners from around the world, the ICC 2024 Group A Code committee voted to allow 100 percent exposed ceilings in Type IV-B construction. Results from this decision were released in January 2022. This update will be included in the 2024 IBC.

5.4.7 AUTHORITATIVE SOURCES

Table 5.2 lists various authoritative sources referenced throughout Chapter 5 and where they can be found for further research. Many of these must be purchased. However, acquiring up-to-date versions of these guides and standards will ensure the user has access to complete and current information.

---

29   https://www.ri.se/en/what-we-do/projects/fire-safe-implementation-of-mass-timber-in-tall-buildings

STANDARD	WEBSITE
National Building Code of Canada Fire Safety Design in Buildings	http://cwc.ca/design-with-wood/building-code/
AIBC Encapsulated Mass Timber Construction up to 12 Stories	https://aibc.ca/
Mass Timber Resources, Varied	https://www.woodworks.org/design-and-tools/building-systems/mass-timberclt-code-related/
NDS for Wood Construction; NDS Supplement; Special Design Provisions for Wind; and Seismic Manual for Engineered Wood Construction	https://awc.org/codes-standards/publications/nds-2018
Nail-Laminated Timber Design and Construction Guide	https://www.thinkwood.com/products-and-systems/nail-laminated-timber
CLT Handbook-US Edition Design and Cost Optimization Checklists and Downloads	https://info.thinkwood.com/clt-handbook https://info.thinkwood.com/mass-timber-direct-2
CLT Handbook-Canadian Edition	clt.fpinnovations.ca
ANSI/APA PRG 320: Standard for Performance-Rated Cross-Laminated Timber; Glulam Product Guide; Glued-Laminated Beam Design Tables; ANSI/APA A190.1: Standard for Wood Products-Structural Glued-Laminated Timber; ANSI 405: Standard for Adhesives for Use in Structural Glued-Laminated Timber; Many more	https://www.apawood.org/resource-library
American Institute of Timber Construction: Test Methods for Structural Glued-Laminated Timber	https://www.aitc-glulam.org
CSA Standard O177-06: Qualification Code for Manufacturers of Structural Glued-Laminated Timber	https://www.csagroup.org
International Building Code	https://www.iccsafe.org

TABLE 5.2: AUTHORITATIVE SOURCES

Source: Hemsworth Architecture; Image Credit: EMA Peter Photography

CASE STUDY: BRITISH COLUMBIA

NEW BC INITIATIVES SUPPORT MASS TIMBER

GOVERNMENTS AROUND THE WORLD are beginning to recognize the impact of the built environment (i.e., construction) on carbon emissions and to support mass timber as a significant tool to help fight climate change.

The Canadian province of British Columbia, a major player in the North American forest products industry, is taking perhaps the strongest government-sponsored action yet to support and promote mass timber through the creation of the Office of Mass Timber Innovation in the Ministry of Jobs, Economic Recovery, and Innovation.

Forestry is a key part of BC's economy. In 2020, forest products represented 29 percent of BC's total exports, valued at $11.5 billion. Forestry also provides well-paying jobs to more than 50,000 workers.

Mass timber is clearly viewed as a significant part of BC's future. Recent government action includes the following:

- Formation of the Mass Timber Advisory Council, a group of experts from urban planning and development, First Nations, the forest products industry, environmental nonprofits, academia, and local governments. The council will provide advice and guidance toward establishing BC as a leader in the production and use of mass timber, as well as the creation of a mass timber action plan.

- The provincial government has thus far invested $4.2 million in its Mass Timber Demonstration Program. The first round of the initiative funded five mass timber projects and four research projects.

- In 2019, the province adopted new national building code provisions for 12-story mass timber projects in interested communities. Twenty communities have opted in, and the City of Vancouver has also adopted similar provisions in its building bylaws.

Source: Ed White Photographics

- The province has committed itself to the effort to construct government-funded buildings with mass timber to the greatest extent possible. The Royal BC Museum's new collections and research facility in Colwood and the new BC Institute of Technology (BCIT) and Okanagan College student housing will use mass timber construction.

"The face of construction is about to change in British Columbia through our government's investments in mass timber," said Ravi Kahlon, minister of jobs, economic recovery and innovation. "By encouraging greater use of mass timber, we are supporting jobs in research, design, engineering, construction, and forestry using BC expertise and materials."

Coming soon as part of the CleanBC Roadmap to 2030 are three additional mass timber-related initiatives:

- **The Mass Timber Action Plan:** The plan takes a triple bottom-line approach to the mass timber industry, simultaneously keeping people, climate, and economic growth in focus during the pursuit of its primary mission—to accelerate the transition of mass timber from niche to mainstream.

- **The Low Carbon Building Materials Strategy:** It will advance the use of mass timber, wood-based insulation, carbon-absorbing concrete, and concrete made with lignin fibers from trees and other plants.

- **The Circular Economy Strategy:** It will support the province's climate goals and economy. Solving the problem of construction waste and diverting materials from landfills also serves to lower the embodied carbon of wood construction, as end-of-life impacts are of a Whole-Building Life Cycle Analysis.

"BC is well positioned to be the world leader in sustainable design and construction innovations that can transform our economy and positively support our environment and climate footprint," said Michael Green, architect and Mass Timber Advisory Council member. "I believe that mass timber will reshape our skylines with beautiful buildings that respond to the needs of our communities and planet. We are proud of the initiatives of the province and peers as we move new ideas forward together." ◗

CHAPTER 6: BUILDERS

IMPACTS OF THE MASS TIMBER EFFECT

- Embodied carbon will account for 72 percent of all carbon dioxide emissions associated with buildings built in the next 10 years.[1]

- Sustainably sourced wood does not necessarily come at a premium, but sources should be vetted before purchase to be compatible with project and industry carbon goals.

- Wood that is renewably sourced can also store rather than emit carbon, contributing to net-zero carbon construction outcomes.

- Collaborative design processes bring designers, builders, and manufacturers together in a scenario that can more closely control sourcing, waste, and embodied carbon emissions.

- Products sourced from rural areas and erected largely in urban centers bridge the urban/rural divide. When sustainably harvested, mass timber products are widely supported and endorsed by diverse communities.

- Building practices that minimize waste, such as modular mass timber and prefabricated components, are often also associated with improved and more diverse working conditions, contributing to the equity and social sustainability of communities.

Mass timber is a disruptive technology with respect to building construction, with increased off-site fabrication and new, highly collaborative construction approaches. The necessity of these approaches is a result of the size and weight of the structural components, precluding significant modifications at the construction site. As such, many contractors will find the information in Chapter 5 is relevant to builders as teams become more integrated, optimizing the design, schedule, and costs together in real time.

Prefabricated panels are produced once the project's architectural and engineering design is complete and construction site logistics are understood and integrated. This quite naturally creates a high premium on integration of the design, manufacturing, and construction aspects of the project, relying on tight collaboration of all parties from the beginning. The situation incentivizes vertical integration of companies along the supply chain and discourages production of commodity blank panels not assigned to any specific project.

We start this chapter with an overview of data for the whole US building market to provide context for where potential growth might occur. Then, we review each common wood construction style to help readers understand how mass timber fits with other wood construction methods. The third section dives into the details of how to approach and execute successful mass timber projects.

---

1 *Architect Magazine*, The Carbon Issue, January 2020, guest edited by Architecture 2030.

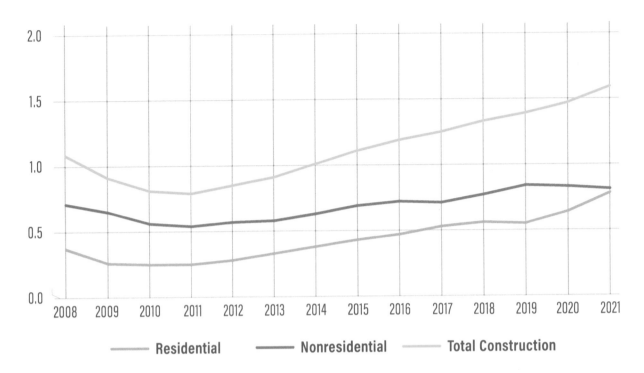

Residential —— **Nonresidential** —— **Total Construction**

TABLE 6.1: ANNUAL VALUE OF ALL CONSTRUCTION, 2008 TO 2019 ($ IN TRILLIONS)

6.1 MARKET CONTEXT

Construction Value

Table 6.1 shows the value of all construction in the United States, per US Census Bureau data. The data is categorized by building use as either nonresidential or residential. The annual value of all construction was over $1 trillion in 2008. It dropped significantly during the Great Recession, but it has since climbed back to nearly $1.6 trillion in 2021. The total increase in value of all construction jumped from over 5 percent in 2020 to over 8 percent in 2021. While residential construction has typically accounted for 30 percent to 40 percent of total US construction values in the last 14 years, it has seen a significant jump to nearly 50 percent during the COVID-19 pandemic.

6.2 WOOD AS A CONSTRUCTION MATERIAL

Compared to the two other primary construction materials, steel and concrete, wood is uniquely strong in both tension and compression for its weight, and it is the only innately renewable structural material with a significant market presence. The structural properties of timber come from trees evolving to carry substantial crown loads and to resist swinging in the wind. These properties make wood highly effective for dynamic loading and fatigue, as well. As such, wood has a high potential for resilience—uncompromised recovery—as a structural material under strong gravity loads, as well as seismic and wind loads. Three types of wood construction are reviewed here: light frame, traditional heavy timber, and mass timber.

FIGURE 6.1: LIGHT FRAME WOOD BUILDING
Source: APA – The Engineered Wood Association

Light Frame

This type of construction, also known as "stick frame," is the most common construction method used for residential buildings in North America. It is also widely used in low- and mid-rise commercial buildings. For lateral resistance and spanning between "sticks," plywood or Oriented Strand Board (OSB) sheathing is commonly used (see **Figure 6.1**).

The advantages of this building system are low cost and ease and speed of assembly. Lumber, plywood, OSB, and other wooden building materials are readily available commodities, and are relatively inexpensive. Additionally, workers can move the building materials around a job site with relative ease compared to larger and bulkier materials such as steel beams, and they can use relatively inexpensive, lightweight tools. Lumber and structural panels arrive on the construction site in bulk, and standard sizes are cut to fit the design at the site. This also means that the design may continue to be adjusted until the framing is complete. All these factors contribute to the widespread use of this construction type for buildings that have lower requirements for fire resistance.

A disadvantage of light frame wood construction is the amount of waste generated on-site, increasing the cost of in-place materials. Of the building

FIGURE 6.2: POST AND BEAM BUILDING
Source: Nordic Structures

styles discussed here, light frame wood carries the highest risk of fire damage.

Heavy Timber

Heavy timber is another traditional method of wood construction, often referred to as "post and beam." In this construction style, large timbers form vertical columns and horizontal beams are connected either with wooden joinery or metal connectors. A key implication of this design is that the columns bear all of the building's weight, meaning the walls are not load-bearing (see **Figure 6.2**).

Because the timber columns and beams bear a building's weight, post and beam construction offers greater design flexibility and allows highly customized and open floor plans. Another advan-

tage is quick completion of a building's structure. Many post and beam designs leave the large dimension beams and columns exposed, for many consumers find the natural warmth and elegance of exposed wood surfaces appealing. In addition, the massive size of the timbers used in a post and beam building provides fire resistance.

Mass Timber

Mass timber (see **Figure 6.3**) refers to engineered wood members that offer a high level of fire-resistance because of their massive size. (See Chapter 1 for definitions of the many types of components that fall into this category.) Up to this point, most mass timber buildings in North America have been low- to mid-rise. However, US building code changes enacted by the International Code Council (ICC) have been incorporated into the 2021 Inter-

FIGURE 6.3: A CLT AND GLULAM MASS TIMBER OFFICE BUILDING

First Tech Federal Credit Union, Hillsboro, OR; Source: Swinerton Builders

national Building Code (IBC) to include buildings up to 18 stories (270 feet). Canada has also developed tall wood code provisions. For more detail and for information on regional adoptions of these codes, see Chapter 5. The benefits and challenges of mass timber construction are explored in detail in the remainder of this chapter.

6.3 THE MASS TIMBER BUILDING EXPERIENCE

When mass timber started making headway as a building material in North America, few building contractors were experienced in expanding what had been the post and beam, heavy timber paradigm. New mass timber technologies necessitated a refreshed approach. This section discusses how

contractors have adapted to using mass timber as a building material and some of the lessons they've learned.

6.3.1 BIDDING AND PLANNING MASS TIMBER PROJECTS

Educating building contractors about the best practices for planning and bidding a mass timber building is an identified industry need. A 2017 report[2] by the British Columbia Construction Association (BCCA), for example, identified barriers to innovation as they relate to using mass timber in buildings. Many barriers were identified, including:

- lack of transparency in the procurement process

- issues over responsibility and allocation of risk

---

2   British Columbia Construction Association, *Procuring Innovation in Construction: A Review of Models, Processes, and Practices*, https://www.naturallywood.com/sites/default/files/documents/resources/procuring_innovation.pdf.

- lack of clear leadership to ensure that construction is properly planned using a design-led approach

- procurement models that inadvertently promote an adversarial relationship among parties

- building contractors who may not be familiar with best practices for managing and mitigating such risks as they pertain to mass timber; when working with mass timber, contract documents should have provisions about weather protection, lifting and storing materials, and fire protection during construction

Each of these barriers are related to a paradigm shift in construction that enforces integration of the design, manufacturing, and construction process. Integration challenges traditional procurement processes and standard allocation of risks and responsibility. A panel manufacturer may include some project management, design, and construction in their scope of work, for instance. Highly integrated approaches offering solutions akin to a turnkey package may create conflicts with parties trying to bid for individual aspects of the project.

Because of the urgent need to train construction teams in the US, WoodWorks has published resources to help guide contractors on the particularities of bidding, planning, and constructing with mass timber. Their "Mass Timber Construction Manual" came out in October 2021. With partners nationwide, Woodworks also offers workshops on mass timber project management and installation training.

To approach bidding and planning mass timber projects successfully, a build team should familiarize themselves with the key considerations covered in this section: optimization during design, procurement, digital tools like Building Informa-tion Modeling (BIM) and Computer Numerical Control (CNC) technology, and prefabrication.

Optimize during Design

A custom mass timber package can save significant field costs, but the benefits are realized only if the design and procurement/build teams work together as early as possible in the design process. Traditional procurement processes are a barrier to early collaboration among designers, builders, and manufacturers. A building owner considering a mass timber building should first be advised on how to choose a procurement process that supports the close collaboration required for the best value outcome (also see Chapter 8). In some cases, the realization of the advantages of integrated mass timber "building production" processes has led to the vertical integration of many traditionally separate functions in one company. An alternative is an alliance of partners with know-how and a history of collaboration.

Each mass timber manufacturer has specific efficiencies and limitations that should be worked into design and logistics plans. Optimization of a structure's design and erection process is what balances out the premium costs of early planning, higher-unit-cost materials, and prefabrication. Early communication among the design, manufacturing, and construction teams can also lead to efficiencies offered by available component sizes, prefabrication, and high-precision CNC finishing. If layout and detail optimization are offered later in the process, such as during bidding, significant redesign may be required to achieve an on-budget package. Pushing design work into the construction phase creates cost and schedule risks; one of the biggest cost advantages of the mass timber construction approach is a dramatic reduction of these risks.

Forest Service
U.S. DEPARTMENT OF AGRICULTURE

Expanding Mass Timber Market Opportunities

Talk to us about getting your mass timber project from the drawing board to market. Our Wood Innovations team manages grants and provides technical expertise to assist in realizing the full potential of the growing demand for mass timber construction. Our Forest Products Lab provides wood product research. The USDA Forest Service is here for you.

USDA Forest Service Wood Innovations Program and Forest Products Lab

- Innovation Grants support design and engineering, development of U.S.-sourced timber, and U.S. manufacturing of mass timber and other wood products
- Research on Building and Fire Science, Sustainability and Life Cycle Assessments

Contact Us
fs.usda.gov
fpl.fs.fed.us

🐦 @forestservice
📘 @USForestService
🐦 @fsWoodLab

Photo Credits (from top to bottom)
Vaagen Timbers
Albina Yard: Jeremy Bittermann courtesy LEVER Architecture
CLT Fire Testing: USFS
Building construction: LEVER Architecture
After harvest: Vaagen Timbers

A successful cost model is necessary to begin construction, but the benefits of early coordination go far beyond cost estimating. Efficient field coordination is where schedule benefits are realized, and a savvy contractor will amplify the structural coordination benefits into other trades as well. For example, a high level of coordination during design was an essential part of the construction-phase success of the 8-story mass timber building, Carbon12, in Portland, Oregon. The project team chose a design-build approach, allowing significant time for Mechanical, Electrical, Plumbing and Fire system (MEPF) coordination with the CLT package. Along with optimizing the structure, the MEPF penetrations were also reduced by careful consideration from an installation-sequencing standpoint. A sequencing plan ensured trades were not in conflict during installation, leading to the subcontractors "working together like a well-oiled machine."[3]

This high level of coordination and "early involvement" of integrated design, fabrication, and construction teams are often offered as a part of a package by seasoned mass timber panel companies in Europe, Australia, and New Zealand and/or by contracting companies specializing in mass timber panel construction as part of the package.

For best practices for early coordination, WoodWorks has created a resource, Mass Timber Cost and Design Optimization Checklists, to assist project teams.[4]

Procurement

The advantages of contractor involvement in project planning include adding valuable insight into material availability. The number of mass timber manufacturing facilities in North America is increasing every year, but available capacity can still vary greatly depending on regional project demands. This supply and demand pressure will continue to shift as the market matures, more facilities come online, and mass timber building designs become more common. Establishing a rough timeline with a manufacturer well in advance of breaking ground will ensure a project meets delivery expectations and avoids the high cost of storing complete sets of massive elements between fabrication and construction. One often-overlooked aspect driving lead time is the custom detailing work needed at the manufacturer during production. Selecting and engaging with a manufacturer early on can help ensure that the team has plenty of time to coordinate and approve shop drawings.

While engineered mass timber components are custom products, they are composed of wood fiber that is subject to the fluctuations of a commodity market. Wood fiber prices can change from month to month, or even week to week, and this plays a part in estimating and timing orders.

BIM and CNC

Mass timber and BIM (see Chapter 5 for more information) are coming of age together, a synergy contributing to the exponential uptake of mass timber technologies. The planning and coordination required for reducing on-site construction time through prefabrication is well supported by a collaborative virtual building model. BIM's potential to streamline coordination through design, manufacturing, and construction is developing rapidly.

---

3 www.buildingCarbon12.com
4 https://www.woodworks.org/design-and-tools/building-systems/mass-timberclt-code-related/

Integrated procurement models are also becoming more common. Procurement barriers discussed in other chapters can limit early coordination for nonintegrated teams, but BIM is also a relatively new technology, and all parties involved are still becoming accustomed to an integrated modeling process. A traditional building contract can also benefit from BIM at all stages.

Using BIM to coordinate a mass timber project can be as basic as the timber manufacturer modeling components for the CNC machine that will cut each panel to precise specifications. However, the process often reaches higher levels of sophistication and can involve each member of the design and build team, depending on the skills of the team and the objectives of the project. Possibilities include detailing down to the level of fasteners, using the model for materials takeoffs and ordering, clash detection for all building systems, and modeling for prefabrication of each building component. This up-front coordination precludes major adjustments of massive components on the construction site, when adding even small cuts to address unforeseen conflicts with other systems can be laborious and costly.

The most common and effective ways to utilize BIM for mass timber are for the architectural, structural, and MEP designers to create intersecting 3-dimensional (3D) models for coordination both in design and construction. These 3D models can also be shared with the mass timber manufacturer for direct use in creating shop drawings for fabrication. Leading mass timber companies and specialized construction companies working with mass timber panels also use BIM for coordinating construction and optimizing sequencing of construction and building finishing jobs.

Prefabrication

Successful projects that maximize prefabrication are pushing the building industry to reconsider project delivery. The American Institute of Architects (AIA) estimates that modular construction projects reduce construction schedules by 30 percent to 50 percent.[5] Modularizing an entire structural system has benefits for on-site safety, schedule efficiencies, and precision, appealing broadly to installers, building owners, and designers. The confluence of BIM and mass timber is leading to increasing conversations about the potential of fabricating more—and more complex—components off-site. In this way, mass timber has become a catalyst for prefabrication in North America, following successful and diverse European precedents.

The potential for off-site fabrication is huge, but facilities are limited in North America. The most common approach is component-based, where large, complex, precise elements are manufactured off-site and set immediately in place, reducing off- and on-site buffer storage needs, installation time, and overall schedules. Flat pack wall systems and volumetric strategies seek to install multiple interacting materials, utilities, and finishes in a climate-controlled interior environment. The benefits include a higher level of quality control and very fast on-site erection times. Whatever the approach, local jurisdictional inspection requirements, transportation limitations, and shipping and handling expenses should be taken into account when strategizing prefabricated building elements. These added costs should be weighed against the potential efficiencies of the design, fabrication, construction, and, at times, potential aesthetic impact of the building.

---

5   *Design for Modular Construction: An Introduction for Architects.*

FIGURE 6.4: PREFABRICATED FACADE PANELS FOLLOW CLOSELY BEHIND STRUCTURAL FRAMING
Brock Commons, University of British Columbia; Source: Ralph Austin at Seagate Structures

As is typical of mass timber, large-scale timber components arrive on site in stacks organized for rapid erection of walls and floors. Because a crane is necessary to move these large components into place, it makes sense to take advantage of the investment and look for opportunities where other time-consuming building elements can be fabricated into larger components, such as facades or mechanical systems. This is especially true for sites where transportation and labor costs are high or lay-down and staging space is minimal, such as remote locations or constrained urban sites.

When MEPF penetrations are precisely located, as with a coordinated BIM process, many components can be fabricated off-site and installed directly in place. Improved planning results in fewer trade conflicts on-site, whether or not additional off-site construction is part of those trades' strategy. But maximizing prefabrication can also lead to a rapid sequencing that is able to keep up with, and take advantage of, the speed of mass timber structural erection.

The 18-story student residence hall, Brock Commons, at the University of British Columbia (UBC) in Vancouver, was erected at 2 floors per week, following the concrete foundation and cores. The Cross-Laminated Timber (CLT) and glulam levels were closely followed by a panelized timber facade (see **Figure 6.4**), providing immediate weather protection and savings in scaffolding, time, labor, and

FIGURE 6.5: PRECISION COMPONENTS QUICKLY ASSEMBLED ON A CONSTRAINED SITE

Project One, San Francisco, CA, Gurnet Point Construction, DCI Engineers, Freres Lumber, Co.

FIGURE 6.6: ASSEMBLING PREFABRICATED COMPONENTS IN A FACTORY SETTING

Source: Katerra; Photo Credit: Kristopher Grunert

risk on-site. In fall 2017, only 66 days from the first panels arriving on-site, the building was structurally topped out and enclosed.

Prefabrication and a design-build partnership were key to the significant schedule savings realized at the 4-story residential building Project One, in San Francisco (see **Figure 6.5**). Located on a very constrained site with no lay-down area, the original structural framing schedule was estimated at 3 months. Using precision-fabricated Mass Plywood Panel (MPP) components from Freres Lumber Co. Inc. for the floors and roof, and panelized light framed walls and moment frames, the structure was completed on budget in just 24 working days. The design-build team worked closely with Freres on design coordination and delivery, and the owner deemed the approach a huge success.

A modular building approach naturally leads to less time on-site, cutting down on local disruptions associated with construction, like increased traffic, lane closures, and noise. Smaller crews require fewer parking spaces, while reduced or eliminated

field modifications make for a very quiet site. Large structural components delivered in predetermined sequences can be off-loaded relatively quickly and immediately set in place, with fewer overall deliveries. In Europe, where urban site constraints frequently have high impacts on construction approaches, mass timber has been found to reduce structural site deliveries by as much as 80 percent. Less lay-down space is needed when installation coincides with just-in-time delivery, another benefit for constrained or sensitive sites.

6.3.2 RELOCATION OF LABOR

Increased prefabrication of building components has excellent implications for the workforce. When more labor takes place at a manufacturing facility (see **Figure 6.6**), on-site construction crews become smaller (see **Figure 6.7**). In a study of 100 mass timber buildings in the United Kingdom, Waugh Thistleton Architects found a 50 percent to 70 percent reduction in site staff for structural framing. In Oregon, the 38,000-square-foot Carbon12 required only 4 carpenters for the 10-week duration of structural erection for all 8 stories.

FIGURE 6.7: A SMALL FRAMING CREW GUIDES PANEL PLACEMENT

Source: The Canyons; Photo Credit: Marcus Kauffman, Oregon Department of Forestry

Factory environments have health and safety benefits for workers when compared to construction sites.

Safety

In a factory setting, there is a dramatic reduction of the hazards experienced on a construction site. Worker safety is improved, and the likelihood of accidents decreases by about half. According to research from University of Utah, "By moving to prefabrication, the construction industry and its workers can experience a much safer environment by a factor of 2."[6]

Climate Control

In some climates, harsh conditions are not only challenging for human health but also limit hours available for construction. For example, a framing crew working in a hot climate will arrive on-site as early in the day as possible to avoid noon sun

6   Ryan E. Smith, *Prefab Architecture* (2010), 86.

exposure, possibly conflicting with local noise ordinances. Prolonged exposure to extreme conditions, as on an unshaded or freezing job site, is stressful to human health and increases safety risks. Controlled temperatures, air quality, noise, and light levels can be provided in an interior environment. Such conditions are healthier and safer for long-term work, and they open jobs up to more candidates.

Commute

Construction workers who commute to a job site are at the mercy of the project location and its distance from their home and community. Some remote job sites require temporary accommodations, and laborers travel home only for weekends. Long and always changing commutes are challenging for families and for an individual's health, and often workers must sacrifice family time, sleep, or other healthy habits.

Ergonomics

For repetitive tasks, a factory can provide more ergonomically designed support. For example, a work surface can be set at a comfortable height for tasks that might require kneeling on-site.

Diversity

Because of the reasons cited above, factory environments provide increased accessibility to jobs for more women, people with health concerns or disabilities, and older workers. Diversity within a company has many proven benefits, including increased productivity, creativity, engagement, profit, and reduced turnover. The benefits ripple beyond projects and companies into healthier, more sustainable communities.

Skills and Training

In a factory producing complex building components, there are opportunities for a wide range of skill sets. A mass timber manufacturing facility will have positions that require little training, as well as positions that require high-level skills and have more earning potential. Unskilled workers are more easily supervised and represent less risk in a controlled facility than on a construction site. Skilled labor might range from craft and finish work to operating computer-aided equipment like a CNC machine or coordinating BIM processes with external design teams. "[T]he prefabrication architecture laborer is much more skilled than any mass-production laborer in previous generations, moving to more intellectual, computer, or even management tasks."[7] Such a range of job opportunities supports diverse communities—especially beneficial in rural communities with limited job options.

6.3.3 PRECISION AND CONNECTIONS

Custom, engineered timber components are very precise, with tolerances in the range of $\frac{1}{16}$ inch. They must be fully coordinated in advance to ensure no field modifications are necessary. Interfaces between mass timber components and other building materials should be identified and proper tolerances allowed for in the design details. Designers should identify where greater levels of precision are most crucial, and contractors can advise on where constructability issues may arise.

7   Ryan E. Smith, *Prefab Architecture* (2010), 87.

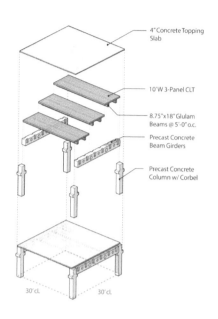

4" Concrete Topping Slab

10'W 3-Panel CLT

8.75"x18" Glulam Beams @ 5'-0" o.c.

Precast Concrete Beam Girders

Precast Concrete Column w/ Corbel

30' cL 30' cL

ABOVE AND RIGHT — FIGURE 6.8: COMPOSITE TIMBER AND PRECAST CONCRETE STRUCTURE

Adidas North American Headquarters, Portland, OR

Source: Lever Architecture and Turner Construction

Installation conflicts can be reduced or eliminated through close, advance coordination of fabrication. Constructability analyses for tolerances are especially important at frequently repeated intersections. A thorough analysis can result in huge risk reductions by avoiding the multiplying effect of repetitive field modifications. Recurring details are also an important opportunity to optimize the sequencing of the build to find schedule and cost savings where possible. Common interfaces, where building in tolerances is crucial to project success, are listed below.

Concrete

Cast-in-place concrete can incur inconsistencies up to 1 inch in multiple planes. Because foundations are typically cast-in-place, the transition between concrete and other framing materials is a connection point that will occur on almost every mass tim-

ber project. Concrete shear walls likewise may have variances from floor to floor, or across a face. A general contractor should impress upon the concrete team to take special care in areas requiring more precision and also flag details that may not allow room for industry-standard installation practices.

Precast concrete is more precise than cast-in-place concrete. This prefabricated solution is worth considering for exposed components with a high level of finish quality (see **Figure 6.8**).

FIGURE 6.9: CLT WALL AND ROOF PANELS WITH STEEL FRAMING

Lincoln City Police Department, OR
Source: Swinerton Builders

FIGURE 6.10: CLT FLOOR DECKS WITH STEEL FRAMING

Brentwood Public Library, Brentwood, CA
Source: Holmes Structures
Photo Credit: Blake Marvin Photography

Steel

Structural steel columns, beams, and braced frames have tolerances greater than engineered wood, typically about ¼ inch to ⅜ inch, and, depending on the length of the steel, up to ¾ inch[8] (see **Figures 6.9** and **6.10**).

The design and fabrication method of exposed or concealed steel connectors, especially details that occur frequently, can significantly impact the schedule of a project. Rolled steel connections will require more tolerance, and it may be wise to plan for shims or other field modifications as needed. As with larger structural components, greater length brings more potential for variation. Highly accurate cast-steel connections may have a higher up-front cost, but they may contribute to schedule savings by reducing field conflicts and retrofits (see **Figures 6.11** and **6.12**).

Rated Connections

Options for achieving required fire resistance ratings, where material tolerances may create gaps at floors, walls, shafts, and other structural connections, should also be evaluated for aesthetics, cost, and constructability.

Carbon12 is an 8-story, hybrid CLT, glulam, and steel-braced frame building with custom steel floor-to-floor connections, and specialized die-cast steel beam-to-column timber connections. The design-build-owner team was under one roof and able to coordinate holistically in preconstruction. The construction manager with Kaiser+Path noted: "In my 30 years of building, I have not seen a building framed as quickly and efficiently as Carbon12. The structural steel core and mass timber elements fit together seamlessly—with very little corrective work."

8   American Institute of Steel Construction.

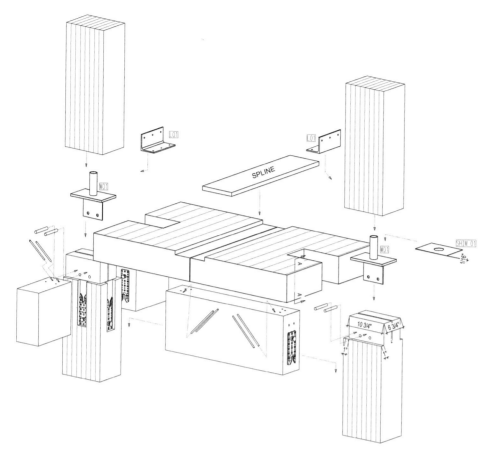

TOP — FIGURE 6.11: KIT-OF-PARTS ASSEMBLY DIAGRAM
FOR TIMBER COLUMN, BEAM, AND CLT FLOOR ATTACHMENTS

Carbon12, Portland, OR; Source: Kaiser + Path

BOTTOM — FIGURE 6.12: OFF-THE-SHELF BEAM CONNECTIONS
AND CUSTOM COLUMN CONNECTIONS

Carbon12, Portland, OR; Source: Kaiser + Path

Source: StructureCraft; Photo Credit: University of Idaho Photo Services

A VISIONARY DESIGN CREATES AN UNUSUAL ARENA

CASE STUDY: IDAHO CENTRAL CREDIT UNION ARENA

STORY CREDIT: OPSIS ARCHITECTURE AND STRUCTURECRAFT

LOCATION: MOSCOW, IDAHO

COMPLETION DATE: FALL 2021

OWNER: UNIVERSITY OF IDAHO

ARCHITECT: OPSIS ARCHITECTURE

STRUCTURAL ENGINEER: STRUCTURECRAFT (ROOF STRUCTURE)/ KPFF (BASE BUILDING)

CONTRACTOR: HOFFMAN CONSTRUCTION

MT MANUFACTURERS: STRUCTURECRAFT, BOISE CASCADE, QB CORPORATION

MT INSTALLATION: STRUCTURECRAFT

THE DREAM OF A new basketball arena at the University of Idaho has been 50 years in the making. The university's vision of an iconic mass timber structure intertwined with the state's natural resources served as a genesis for the project. Situated next to the historic Kibbie Dome, the new Idaho Central Credit Union Arena establishes a dramatic gateway to the University of Idaho campus through its expression of innovative and sustainable mass timber design.

Opsis Architecture envisioned a landmark building complementing its environment. The undulating roof forms recall the surrounding Palouse landscape.

Source: StructureCraft;
Photo Credit: University of Idaho Photo Services

Source: StructureCraft
Photo Credit: University of Idaho Photo Services

Every aspect of the arena enhances sustainability and the spectator experience, and it showcases the region's characteristics. Structural engineer, timber fabricator, and installer StructureCraft was integrally involved in concept design with Opsis, shaping the curved roof and optimizing it for fabrication and constructability. The roof is a doubly curved plywood diaphragm, acting as a deep beam for seismic loads and allowing the roof to span 360 feet without requiring interior bracing. A portal frame that spans 120 feet allows unobstructed views of the court from the upper-level seating area. Complex timber engineering was required to design the thrust connection between beam and column to transfer over 450,000 pounds of compression.

Three-dimensional modeling yielded all the geometry and shop drawings required to produce the curved glulam shapes. StructureCraft's skilled shop team used Computer Numerical Control (CNC) machining along with layout optimization to cut the desired shapes from the curved billets received from Idaho manufacturers.

The general contractor, Hoffman Construction, facilitated and provided oversight of the installation, working with timber installers from Structu-

Source: StructureCraft

reCraft to click together the massive prefabricated components.

The 4,000-seat arena is set to be a model for the use of timber in long-span sports facilities. Mass timber is a sustainable, future-thinking construction type. It brings together owners, educators, designers, and suppliers to create beautiful spaces people love to be in. ◐

FIGURE 6.13: TIMBER CONSTRUCTION ON CONSTRAINED URBAN SITES

(LEFT) *Sideyard, Portland, OR; Contractor: Andersen Construction Source: Catena Engineers; Photo Credit: Skylab Architecture*

(RIGHT) *District Office, Portland, OR; Source: Andersen Construction*

6.3.4 ON-SITE MATERIAL MANAGEMENT

Perhaps the most important lesson from the first mass timber projects developed in North America is that on-site material management is crucial for efficient construction. The following topics outline the advantages and challenges specific to handling mass timber components on a job site.

Just-in-Time Delivery

In situations where on-site storage is limited, mass timber panels can be delivered on flatbed trucks using a just-in-time delivery system. Such a system takes considerable planning and coordination with both the trucking company and the mass timber manufacturer. Many mass timber companies use their own trucking precisely for that reason. A side motive is that loading in construction sequences may take more time than conventional trucking companies may be willing to tolerate without extra charges. Another is the care for the integrity of weatherproof packing,

which may need some attention and adjustments during transportation in inclement weather.

The just-in-time approach can be complicated by greater distances between the building site and the mass timber manufacturer, regional restrictions on oversized loads, routes with clearance constraints, challenging terrain, or constrained urban sites (see **Figure 6.13**). Unusually shaped panels are more challenging to balance for transport, potentially increasing the number of trucks required or complicating sequencing. The transport team can advise on route strategies and restrictions and any added costs associated with oversized loads.

The challenges of materials management within a given space at a building site aren't specific to mass timber. However, each prefabricated mass timber element has a precise installed location, which has additional site coordination implications. When panels are loaded for shipping at the manufacturer, they are ideally placed with a consideration of the order in which they will be

FIGURE 6.14: STAGING AND HANDLING
Source: Nordic Structures

placed. This approach allows for smooth off-load sequencing and installation, without the need for on-site storage. Efficient and safe loading of the material on the trucks will often take precedence, and it will also be informed by weight distribution, as well as by panel size and shape. Understanding the loading and shipping approach before the material arrives on-site reduces delivery conflicts. A building design with many similarly sized panels will be more straightforward to coordinate than one with many unusual shapes. In the latter case, some lay-down space should be set aside for re-sequencing (see **Figure 6.14**).

Coordinating a huge volume of mass timber materials has storage, schedule, and liability implications at both the manufacturing facility and the construction site. A recent case study published by the DLR Group[9] recommends that the construction team dedicate an engineer to manage

a project's mass timber fabrication and delivery schedule.

Support Equipment

It is important to determine the amount and type of support equipment needed at the site to ensure efficient operation. Some case studies describe using forklifts or similar equipment to move mass timber around the site (only an option in 1- or 2-story buildings) versus using a crane. If small equipment is to be used, the vehicles must be large enough to carry heavy timbers and panels. For example, a 5-ply, 10-foot-by-60-foot panel made from Douglas fir weighs over 5 tons. If panels arrive in a container, which is common for materials supplied from overseas, the equipment on-site must be robust enough to remove and lift heavy panels and timbers. Additionally, enough space is needed to safely maneuver around the site.

9   DLR Group, *Tall With Timber: A Seattle Mass Timber Tower Case Study* (November 2018), http://www.fastepp. com/wp-content/uploads.

FIGURE 6.15: CRANE LIFTING WITH SLINGS

District Office, Portland, OR
Source: Andersen Construction; Photo Credit: Pete Eckert

FIGURE 6.16: PICK POINT LIFTING DEVICE

Source: https://mtcsolutions.com/ (formerly My-Ti-Con)

Most projects will opt to use cranes. This allows for panels or timbers to be "flown" from a truck or site storage into the designated place in the building, as in **Figure 6.15**. A key aspect of this process is the placement, number, and strength of the "pick points," or lifting devices.

Figure 6.16 illustrates a typical lifting device called a Yoke 1T that has been designed and tested for use in mass timber construction. The device is screwed into a mass timber panel using ½-inch screws and is designed to safely lift panels of up to 7,000 pounds. Other lifting devices are available

that are designed for lighter or heavier panels. A key to efficient construction is placing the lifting devices on the panel in a way that allows it to balance plumb and level, easing installation. The pick points also enhance safety by serving as a place for construction workers to "tie-in" after the panel/timber is in place.

Some companies enhance and simplify lifting by cutting small penetrations that engage with lifting slings or crane fixtures. These strategically positioned penetrations are then plugged at the construction site. Others offer quick-mount/

quick-release fixtures to reduce lifting cycle time. Some are designed to reduce the size and visibility of the permanent mark on the panel.

Waste Management

Because mass timber is prefabricated, there should be little to no field cutting of material, resulting in very little wood waste at the jobsite. Builders report that this contributes to enhanced safety because the site stays clean, and storage and removal of waste doesn't require management's attention.

Panels often come wrapped in plastic for protection during transport and on-site storage. While lightweight, this plastic comprises the bulk of on-site waste volume associated with mass timber, and it is destined for the landfill. There is potential for this waste stream to be reduced or eliminated if the protection can be made reusable or multifunctional.

Metric Units of Measurement

Although the capacity of North American mass timber manufacturers is ramping up, some building projects are utilizing mass timber produced in Europe, where the measurement units are metric, rather than the imperial system used in the US. Several builders who dealt with this issue reported that they (and their carpenters) were initially very worried about the differing units of measurement. Crews were supplied with tape measures showing both imperial and metric measurements. That approach created confusion. The solution reported by all builders was to use tape measures calibrated only in metric units, and the crews quickly adapted.

Source: Bryant Photographics

CASE STUDY: VAPROSHIELD

PRIVATE MASS TIMBER RESIDENCE, COLVILLE, WASHINGTON

VAAGEN TIMBERS BUILT A dream mass timber home for mass timber proponent Tom Baun's family in the beautiful forest town of Colville, Washington, in the Pacific Northwest. VaproShield is proud to be part of this sustainable, dynamic home with an expected completion date of spring 2022.

Question: What are some of the advantages of mass timber construction versus conventional construction?

Tom Baun: The speed of construction is second to none. We were able to erect the frame well ahead of schedule.

Question: What were some of the hurdles you had to overcome to use mass timber?

Baun: Since mass timber is a relatively new product to the United States, inspectors, designers, and contrac-

tors are just getting used to working with the product. Once I was able to educate them about mass timber, hurdles and doubts were quickly removed.

Question: Mass timber is obviously wood, and it rains a lot in the Northwest. Why did you decide to use VaproShield water resistive barrier (WRB) air barrier roof underlayments and wall membranes?

Baun: First, VaproShield membranes were proven to work with mass timber; they have completed numerous large-scale mass timber projects around the US. Next, since VaproShield membranes are "breathable," like Gortex for your house, they wick moisture away from the wood substrate and diffuse the water vapor into the air. This allows the mass timber to continually dry itself out, which is exactly what we need here in

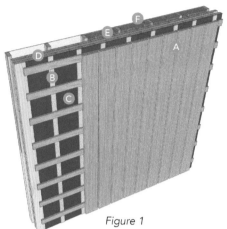

A Vertical cedar siding with 3/4" reveal

B Batten/counter batten rain screen and siding attachment system

C RevealShield SA Self-Adhered Air Barrier WRB membrane for open joint cladding

D 3-ply CLT

E Mineral wood insulation

F Interior wall

Figure 1

Source: Bryant Photographics

the Pacific Northwest. Other products simply didn't offer the high perm rating, and they required additional primers to adhere to the mass timber.

Question: What was the wall assembly buildup?

Baun: You can see from Figure 1 that we included VaproShield's RevealShield SA, which is directly adhered to the mass timber substrate; then added a batten/counterbatten system, creating a rainscreen gap; and finished with cedar planks. On the warm side of the wall, we have wood framing, with mineral wool insulation, finished with drywall. The entire system is optimized to allow the mass timber to dry out through vapor diffusion.

Question: The structure has both a flat and sloped roof. What details did you use for those configurations?

Baun: We had multiple details for the roof assembly, but the common element on the entire roof is Slope-Shield Plus SA by VaproShield, a breathable roofing underlayment. We added various types of insulation over SlopeShield Plus SA and finished with shingles

and thermoplastic polyolefin (TPO) roofing systems. In all instances, the mass timber is exposed on the inside.

Question: You mentioned that mass timber helps the home to be sustainable. What products and efforts are helping to make that happen?

Baun: The native timber was sourced directly from our local fiber basket, within 100 miles of our home and the Vaagen Timbers plant. Plus, both Vapro-Shield products have zero volatile organic compounds (VOCs) and RevealShield SA Self-Adhered, used on the walls, offers the lowest cone calorimeter burn rating of any self-adhered air barrier WRB in the industry. ◖

FIGURE 6.17: TIMBER FRAME AND STEEL CORE PROGRESSING IN COLD, SNOWY WEATHER

Carbon12, Portland, OR; Source: Kaiser + Path

6.3.5 WEATHER PROTECTION AND MOISTURE MANAGEMENT

Mass timber has both inherent advantages and challenges associated with weather. Unlike concrete, which has curing limitations around temperature and precipitation, and steel, which requires certain conditions for proper welding, mass timber components can be installed regardless of weather conditions. This has excellent implications for reducing weather delay contingencies during timelines that overlap challenging weather months.

For example, the framing for Carbon12 took place between December 2016 and February 2017, which was one of the wettest and coldest winters in recent history in Oregon (see **Figure 6.17**). While most of the construction sites in town were closed for several days at a time through the season, this project was only delayed for one day, when key members of the 4-person framing crew were unable to travel to the job site due to road conditions.

Once in place, wood components require some protection against exposure to wet weather to prevent moisture uptake.

Short of coordinating construction around a dry season, only occasionally a viable option, having a moisture management plan in place will help the team manage site practices and invest in protection measures that best fit the project. This plan should be distributed and discussed with all trades that will be on-site during wet weather. Top concerns include staining, swelling, shrinkage, and decay, which can all be avoided by

FIGURE 6.18: DISTRICT OFFICE IMPLEMENTED A MOISTURE MANAGEMENT PLAN

District Office, Portland, OR
Source: Andersen Construction

following a well-considered protection and mitigation plan (see **Figure 6.18**).

Industry standard practices for moisture management in mass timber buildings are developing. In early 2020, RDH Building Sciences published advice on moisture risk management for mass timber builders.[10] Meanwhile, experienced builders are also developing best practices. While constructing both the George W. Peavy Forest Science Center and the District Office during Oregon's wet months, Andersen Construction created a 4-part moisture management plan for wood structures: sealers, stain prevention, moisture control, and

dry-out. Each is elaborated upon below. (For more on managing moisture throughout design and construction, see section 5.2.9.)

Sealers

Shop-applied sealers can protect against moisture intrusion during construction, and some may come standard—or as an option—with some mass timber products. All component surfaces may benefit from different types of sealers, whether applied before delivery or on-site. Facility capabilities vary and should be fully understood if sealers are to be relied upon for weather protection.

---

10 RDH Building Sciences, *Moisture Management for Mass Timber Buildings* (2020).

Source: Kruse Smith Entrepreneur

CASE STUDY: SOPHIE RADICH SCHOOL

LOCATION: LILESTROM, NORWAY

DEVELOPER: SKEDSMKO MUNICIPALLY

CONSTRUCTION: KRUSE SMITH

ARCHITECT: ARKITEMA

HOW CAN THE QUALITY assurance system of the future help mass timber projects? Kruse Smith Entrepreneur is a leading Norwegian project developer and contractor that is building a school in solid timber, in which most of the elements are Cross-Laminated Timber (CLT).

To stay on top of flaws and damage during the construction phase and thereby increase the security of the developer, Kruse Smith incorporated an intelligent sensor solution from the Danish company Woodsense.

The sensors can be placed throughout the project to help monitor particular exposed areas, pick up moisture before damage happens, and document flawless construction all the way from the manufacturer to the construction site.

Through constant monitoring and automatic alarms, the sensors assist in carrying out an effective moisture content strategy by documentating correct moisture management through the entire construction process. The sensors can be used for correct execution of the quality assurance plan and documentation of the level of moisture in the elements.

DETECT AND PREVENT DAMAGE

To monitor the construction of the Sophie Radich school, a sensor was placed on a cornice to monitor the cover toward the edge. Here, the sensor found a mistake made by the roofer, as the covering had not been completed, resulting in water entering the building.

Had the sensor not found the damage in the cornice, a substantially more expensive repair would have been required. The contractor therefore recommends using sensors for quality assurance.

POTENTIAL FOR MOISTURE MANAGEMENT

Elsewhere on the construction site, sensors have been placed in an area where a tarp is being used to keep water away while crews wait for extra material to cover it. The sensors have discovered that water came in, informing construction teams about when and how much water the exposed areas received.

Had Kruse Smith used the sensors earlier, they would have been placed by a window where the developer had concerns. The window solution had been pushed out from the outer wall, and there had been discussion about where the condensation could be expected to land.

DESIGNED TO INCORPORATE NATURE

The school is set to be completed in the summer of 2022, at which time it will accommodate a total of 820 students and employees.

In addition to being built in timber, the school will be shaped like a 4-leaf clover and will have a strong focus on the surrounding environment, through which the students will experience the interaction between the indoors and the outdoors. ⬤

The top surface of a floor panel is more susceptible to standing water, while the bottom face is more likely to be left exposed as a finished surface and need protection with staining. Moisture uptake is quickest at the end-grain, where timber components are the most vulnerable. It is also where components are typically joined together, creating hidden conditions with less air circulation for dry out. Often, for protection during transport and installation, a temporary wax coating will be applied by the manufacturer to edges where end-grain is exposed.

Stain Prevention

Managing construction activity on a mass timber structure intended for finish exposure is critical for preventing stains. Communication is an im-

portant component of a stain prevention plan, as many trades are unaccustomed to working around finished surfaces. Some superficial stains can be cleaned or sanded, but proper stain prevention will avoid the risk of permanent marking, as well as reduce cleanup time and expense. Because multilevel buildings often have repeating floor layouts, penetrations and panel seams can create pathways for water to move from floor to floor. Water readily transports pigments from debris, such as rust from metal-work shavings or other untreated metals, or even a spilled beverage.

Moisture Control

Two basic concepts are paramount to controlling moisture in structural wood. First, protect wood

FIGURE 6.19: CLT PANELS PROTECTED WITH WRAP FOR TRANSPORT AND ON-SITE STORAGE

Hillsboro Community Center, Hillsboro, OR
Source: Swinerton Builders; Photo Credit: BREWSPHOTO LLC

from prolonged exposure to water. Second, if wood becomes wet, it must be allowed to release moisture via proper ventilation.

As soon as mass timber components leave a climate-controlled fabrication facility, they are subject to shifting moisture content, depending on the environment to which they are exposed. Mass timber manufacturers are responsible for protection during transport, commonly accomplished by durable plastic wrap, as shown in **Figure 6.19**. Once the timber is delivered to a project site, the contractor is then responsible for protection, whether stored or in place.

Strategies for protection may be holistic, as in a tented approach, or local, such as tape at panel seams and penetrations.

In Nordic countries, it is quite common to conduct mass timber construction under large-scale tents doubling as an overhead crane support. Variations appropriate to the scale of construc-

tion are available. This approach is controversial because of the added cost and is the subject of a robust industry-wide discussion in the context of the related cost/benefits.

Fully tenting a structure eliminates the need for many of the practices described in this section, but it is often prohibitively expensive, and most projects will need to implement a multipronged approach. Standing water should be minimized and removed as quickly as possible. The construction team should prepare for dewatering activities by having adequate equipment and personnel on-site following rain events, as well as a planned approach for continuous wet weather.

In addition to protection, the basic principles of any approach must allow for wood to release excess moisture at an appropriate rate until the structure has reached equilibrium with ambient environmental moisture during occupancy (see also Chapter 5.1.8 on moisture). Moderate drying conditions and longer drying times help prevent surface checks in elements intended to be visible. Otherwise, such checks in structural elements are of no particular consequence.

Dry Out

Industry standard best practices for acceptable moisture content mass timber have not yet been established. However, in the Pacific Northwest, where wet winters impact construction sites significantly, teams have found that mass timber components that are above about 14 percent moisture content (MC) should not be enclosed or encapsulated, but given a controlled opportunity to release moisture.

Mass timber naturally dries out more slowly than light framing because of its larger dimensions. Because of this greater volume, there is more potential for moisture content differentials within a single panel or member. The greater the differential in moisture, the greater the potential for tension, compression, and movement—created by swelling and shrinking—as the wood takes on water or dries out. These stresses in the wood can lead to cracking and checking, which, while typically structurally insignificant, can be aesthetically undesirable.

6.4 QUANTIFYING COST SAVINGS

There are many reasons that mass timber buildings can be less costly than other construction types. Cost estimating, however, is traditionally based on a wealth of data from past projects, and few contractors in North America have a portfolio of mass timber data to draw from yet. As previous sections have illustrated, early estimates that are not holistically coordinated with the de-

sign, procurement, and logistics teams will very likely be inaccurate.

One of the most quantifiable ways to estimate the difference, and one that will have many ripple effects on cost for the building owner, is through the schedule. Mass timber construction happens much more quickly, with lighter equipment, and with less on-site labor than a comparable building of steel or concrete. The savings are realized on the person- and equipment-time scale; fewer resources are required for a shorter period.

A challenge associated with validating this claim is that there is rarely a case where identical buildings are constructed using different structural materials, thereby allowing an apples-to-apples comparison. There may be cost comparisons between structural materials, but they are based on plans and estimates, not on actual construction costs. In addition, developers may want to test different structural materials for the same project, and then perform a comparative cost analysis. In a process like this, when high unit cost items are flagged for replacement with lower cost materials, mass timber is often eliminated. Looking holistically at estimated schedule impacts is crucial when comparing mass timber with other building materials. Just as important is considering material reductions throughout the building, such as reduced foundations and excavation costs, and the elimination of drywall, framing, and painting of exposed wood surfaces.

Source: Bryant Photographics

CASE STUDY: MONTANA FWP CONFERENCE CENTER ADDITION

MASS TIMBER ON A MINOR SCALE

LOCATION: MISSOULA, MONTANA

COMPLETION DATE: SUMMER 2021

OWNER/DEVELOPER: MONTANA FISH, WILDLIFE AND PARKS

ARCHITECT: MMW ARCHITECTS

MECHANICAL/ELECTRICAL/PLUMBING AND STRUCTURAL ENGINEERING: MORRISON-MAIERLE

CONTRACTOR: QUALITY CONSTRUCTION COMPANY

MASS TIMBER MANUFACTURERS: SMARTLAM NORTH AMERICA

FWP BRINGS MASS TIMBER HOME

FROM ITS INCEPTION, THE Montana Fish, Wildlife and Parks (FWP) Conference Center Addition was seen as a way to demonstrate the use of Cross-Laminated Timber (CLT) in Montana. Although it has one of the few CLT manufacturing facilities in the county, Montana lacks the significant growth in CLT construction that's common in other regions.

Because the addition was the first full CLT project commissioned by the State of Montana, the project was seen as a way of providing exposure to design professionals, contractors, and developers in the surrounding regions. The State of Montana also views CLT as an opportunity to revive the timber industry, which was the state's largest manufacturing sector 30 years ago. With support from the governor's executive order to promote the use of Mon-

Source: Bryant Photographics

Source: Bryant Photographics

tana wood products, all lumber for the CLT panels was harvested and manufactured in Montana.

LEARNING THROUGH PRACTICE

Even though the conference center addition is a relatively small project, those involved took care to explore ways CLT could be used to reduce construction time while increasing accuracy through a prefabricated system. Using an integrated design approach, the architecture and engineering teams engaged with the local CLT manufacturer during design to coordinate key issues. They used Building Information Modeling (BIM) for each panel to better understand the sequence of installation, joint layout, and penetration locations. The designers used Computer Numerical Control (CNC) machines to scribe layout lines and facilitate the assembly of complicated intersections.

This partnership continued through the bidding process, as they collaborated on a workshop for prospective bidders. The workshop offered a crash course on CLT construction and how it applied to the project, in addition to an on-site mockup of a corner-to-roof intersection. The mockup demonstrated the benefits of precut penetrations, precut beveled connections, and simple mechanical connections.

The workshop gave local contractors an opportunity to familiarize themselves with CLT construction, resulting in a positive bid turnout.

The project was awarded to a local construction company. During review of the shop drawings, the entire team worked with the general contractor and subcontractors to analyze the details to maximize the efficiency of the prefabrication process. Particular scrutiny was given to the routing of any exposed distribution systems. In only minor instances did penetrations have to be cut in the field.

Once the CLT panels arrived on-site, the walls and roof were installed in approximately 3 days. The general contractor indicated that the accuracy and speed of construction was a substantial benefit to delivering the project on time during Montana's limited building season.

The Montana FWP Conference Center Addition embodies the department's mission for the stewardship of natural resources for present and future generations, captures the awe-inspiring beauty of wood, and serves as a demonstration of the potential for mass timber construction. ⌂

MTS

Mass Timber Strategy

The Gateway to Mass Timber

Helping you navigate the world of Mass Timber through strategic advice and referrals.

Visit the Mass Timber Marketplace at

www.masstimberstrategy.com/marketplace

Our Partners:

CHAPTER 7: OCCUPANTS

IMPACTS OF THE MASS TIMBER EFFECT

- Demand for comfortable, healthy interior spaces drives the market for sustainably sourced wood buildings.

- Exposed wood surfaces support biophilic responses in building occupants, promoting health and productivity in all building types.

- Unfinished wood has antibacterial, hypoallergenic, and hygroscopic properties that contribute to human health and well-being.

- Spaces that give occupants a "sense of place," such as visible, locally sourced wood, are correlated with environmentally conscious behaviors,[1] multiplying the benefits of a carbon-sequestering wood building.

Indoor environmental quality (IEQ) is a measurement of how a building affects its occupants' comfort and health. An Environmental Protection Agency (EPA) study[2] found that in the US, respondents spent about 87 percent of their time inside buildings and an additional 6 percent in cars, for a total of 93 percent. Canadians fare about the same, at 94 percent, and Europeans spend only slightly less time indoors, 90 percent. The study suggests that people should spend more time outside because a growing body of scientific evidence links interactions with nature to greater levels of health and happiness. It also suggests that interior spaces should incorporate natural elements as much as possible to ensure health.

The powerful influence of nature in all aspects of indoor environments is known as "biophilia," the innate human love for natural forms. Our bodies, as biological organisms, are supported by biophilic spaces.

Mass timber buildings can boost the health, well-being, comfort, productivity, and prosocial behaviors of residents. Human health, comfort, and behavior are closely related, but they are divided into three sections in this chapter. The first section, on health, looks at our acute biological responses to indoor environments; the following section, on comfort, reviews the universal characteristics of those spaces and of human preferences; and, in the section on behavior, we consider how indoor environments influence how we interact with one another.

7.1 HEALTH

The focus on health benefits of wood in the built environment is based on a well-established body of research showing that exposure to nature has health benefits, such as lower blood pressure, lower heart rate, increased ability to focus, increased concentration, and increased creativity.

7.1.1 BIOPHILIA

The idea of enhancing human health through building design has been described as the application of biophilia in the built environment. "Biophilia" is a term that was coined by biolo-

---

1   https://www.stockholmresilience.org/research/research-news/2017-05-30-a-better-sense-of-place.html

2   Neil E. Klepeis, et al., *The National Human Activity Pattern Survey (NHAPS): A Resource for Assessing Exposure to Environmental Pollutants* (2001), https://indoor.lbl.gov/sites/all/files/lbnl-47713.pdf.

gist Edward O. Wilson, a professor emeritus and researcher at Harvard. He defined it as the urge to affiliate with other forms of life in nature. Biophilic design in buildings connects occupants to nature by featuring natural materials, shapes, and patterns; orienting a building to take advantage of daily and seasonal light patterns; and providing views and access to the outdoors and nature.

Some of the most comprehensive data gathered around the benefits of biophilic building design on human health is captured in a document by Terrapin Bright Green, a design consulting firm.[3] According to studies cited in the report, nature-oriented design improves health by lowering stress and blood pressure; improves mental functions, stamina, and focus; improves moods and learning rates; and decreases violent and criminal activity.

7.1.2 STRESS REDUCTION

A study[4] by FPInnovations connected the use of wood to supporting human health in the built environment. The study documented a lowered sympathetic nervous system response when occupants could see more wood surfaces in a mock office environment. Stress, as measured by heart rate and skin conductivity, was lowest for the participants in the office with the wood design. If extended to an entire building, the study suggests that mass timber is well-positioned to enhance the health of a building's occupants.

Another study, by Japanese researchers[5] in 2007, monitored subjects' physiological responses to different ratios of wood surfaces in an environment. They discovered that a moderate ratio (45 percent coverage) was subjectively "comfortable" because it lowered blood pressure and increased pulse rates. A large ratio of wood surfaces (90 percent) "caused significant and large decreases" in the blood pressure of test subjects.

This topic is drawing increased cross-disciplinary interest. Similar projects are being conducted at the University of Helsinki in Finland, in the University of Primorska in Slovenia, and likely by many other academic and private research groups.

7.1.3 RECOVERY AND HEALING

Another emerging area of occupant health is evidence-based design, involving the analysis of the design of a building to assess how it impacts human health. Architects specializing in the design of health-care buildings are using wood to enhance patient recovery and health, and to optimize the well-being of staff and visitors. One study of human response to health-care facilities found that using cedar panels in hospital rooms reduced stress as measured by cortisol levels.[6]

Biophilic design in health-care environments is linked to shorter hospital stays, faster recovery rates, fewer negative comments from hospital staff, and reduced medications.[7]

3   Terrapin Bright Green, "The Economics of Biophilia: Why Designing with Nature in Mind Makes Financial Sense," 2014.

4   FPInnovations, *Wood and Human Health* (2011).

5   Y. Tsunetsugu, Y. Miyazaki, and H. Sato, "Physiological Effects in Humans Induced by the Visual Stimulation of Room Interiors with Different Wood Quantities," *Journal of Wood Science* no. 53 (2007): 11–16.

6   FPInnovations, *Wood as a Restorative Material in Healthcare Environments* (February 2015).

7   Terrapin Bright Green, "The Economics of Biophilia: Why Designing with Nature in Mind Makes Financial Sense," 2014.

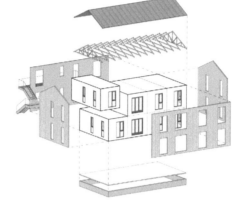

7.1.4 INFECTION CONTROL

The year 2020 brought an increased awareness of how the air and the surfaces around us contribute to our safety or exposure to contagion. An ongoing Finnish study has shown that "the contagiousness of coronaviruses decreases much more rapidly on a wooden surface than on other materials, such as plastics."[8] Wood is an effective antibacterial surface, especially when compared to materials like glass or plastic. Another Finnish study found that pine, spruce, and birch surfaces effectively prevent the growth of pathogenic bacteria common in hospitals, such as the kind that cause staph infections.[9]

The Institute for Health in the Built Environment (IHBE) at the University of Oregon is also doing ongoing research that observes how the unique natural properties of wood could make it difficult for different pathogens to survive or be transferred to occupants on wood surfaces. Wood has a porous surface that can both sequester moisture and desiccate it. Wood contains aromatic organic compounds called terpenes, found in many plants, called terpenes, that appear to have antiviral effects. These IHBE studies are investigating the effect of wood species, coatings, humidity, and simulated flooding events on the surface and air microbiome in exposed wood buildings. Other IHBE studies have shown wood's promise for promoting healthy bacteria and supporting diverse indoor biomes that contribute to human health.

These studies have the potential to significantly increase the use of wood in health-care environments.

7.2 COMFORT

IEQ in relationship to occupant comfort is multidimensional, including thermal comfort, indoor air quality, acoustics, visual comfort, and safety. In the simplest terms, when people feel comfortable in a built environment, they also tend to be more healthy and productive. Mass timber buildings can enhance occupants' comfort in several ways.

7.2.1 INDOOR AIR QUALITY

Many factors contribute to healthy indoor air quality (IAQ) that are beyond the scope of this report, including ventilation rates, filtration systems, outdoor air quality, and occupant behavior. We focus here on providing information about how using exposed wood in interior spaces can support high IAQ characteristics, as part of a complete healthy building system.

7.2.2 TOXICITY

Wood is considered hypoallergenic, meaning it is very unlikely to cause allergic reactions, and its smooth surfaces are easy to keep clean and free of particles. Mass timber panels are manufactured using resins that result in virtually no formaldehyde off-gassing. Many mass timber products are "red-list"-free[10] and approved for use in living buildings.

---

8   Antti Haapala, University of Eastern Finland.

9   Tiina Vainio-Kaila, doctoral thesis, Technical Research Centre of Finland.

10   The RED List contains 22 classes of chemicals prevalent in the building industry, which the International Living Future Institute has designated as worst-in-class.

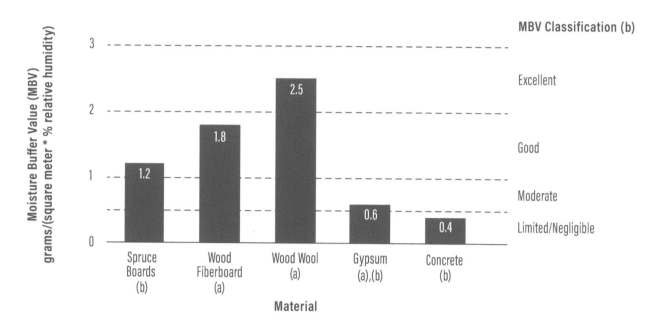

FIGURE 7.1: MOISTURE-BUFFERING VALUES OF COMMON BUILDING MATERIALS

(a) Bruce King et al., "A New Carbon Architecture, referencing Holcroft, N.A. 2016, Natural Fibre Insulation Materials for Retrofit Applications" (PhD thesis, University of Bath, UK).

(b) Rode, Peuhkuri, Time, Svennberg, and Ojanen, Moisture Buffer Value of Building Materials (2006).

Relative Humidity

Relative humidity (RH) is the amount of moisture in the air expressed as a percentage of potential moisture in the air at the same temperature in the environment. The optimum range for human health is 40 percent to 60 percent RH, coinciding with the least optimal range for human health-challenging organisms like bacteria, viruses, fungi, and mites. Just as materials with high thermal mass, like stone or concrete, absorb heat on a sunny day and release it in the cool of night, so, too, can different materials contribute to balancing humidity levels.

Because wood is hygroscopic, it assists in moderating humidity levels by absorbing moisture during periods of high humidity and releasing

moisture during periods of low humidity. The ability of any given material to perform this function is measured by its Moisture Buffering Value (MBV). Values over 1 (g/[m²%RH]) are considered good, and materials with values over 2 are excellent. As illustrated by **Figure 7.1**, wood products perform very well—2 to 5 times better than other tested common indoor materials, including gypsum board and concrete.

7.2.3 ACOUSTICS

Acoustics from an occupant's perspective can be classified in two ways: structure-borne and ambient. Buildings with design features that control for both can significantly enhance occupant satisfaction. Adding mass to an assembly is an important aspect of acoustic mitigation. The

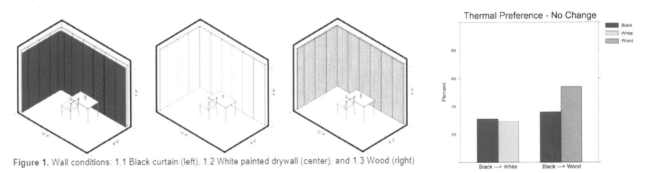

Figure 1. Wall conditions: 1.1 Black curtain (left), 1.2 White painted drywall (center), and 1.3 Wood (right)

FIGURE 7.2: STUDY FINDINGS ON THERMAL COMFORT

Visual effects of wood on thermal perception of interior environments
Denise Blankenberger, Kevin Van Den Wymelenberg, Jason Stenson, University of Oregon, Eugene, OR, 2019

sound-dampening qualities of solid wood have long been recognized, and mass timber performs well in managing structure-borne sound.

In Europe, where noise and vibration transmission standards for various classes of buildings are more stringent than in North America, many panelized mass timber projects use special durable polymer dampening seals in pedestal-type floor-to-wall connections to further reduce vibration and sound transmission.

Ambient sound experience can be managed with sound-absorbing materials to control reverberation of noise in a space. Architectural finishes, furnishings, and even occupants themselves can absorb sound. Wood is a porous material and contributes well to the absorption strategy. It also has an interesting impact on an occupant's perception of noise. A 2019 study[11] at the University of Oregon investigated how wood affects ambient sound comfort by collecting biometric data from building occupants, including measuring galvanic skin response, heart rate, and emotional responses using facial recognition software. They compared masonry and mass timber in office environments, and they found that the exposed wood in mass timber buildings may provide an "acoustic forgiveness factor" when occupants are exposed to similar distracting stimuli throughout the day. That means the same sounds that irritate a person in a masonry building may not have the same negative effect on someone in a space with significant biophilic features, in this case, wood.

7.2.4 THERMAL COMFORT

Wood-framed buildings perform well thermally because wood is a natural insulator. This gives designers increased flexibility when detailing insulation to meet energy efficiency codes, making *actual* thermal comfort a feature of a well-designed wood building. Wood additionally contributes to a *perceived* sense of thermal comfort, broadening acceptable temperature ranges, and saving on operational carbon emissions and energy costs.

---

11 Bain, Montiel, Summers, and Yauk, *Auditory Visual Perception: Acoustic Distractions in Mass Timber Versus Concrete Office Spaces* (2019).

FIGURE 7.3: FLOOR-TO-STRUCTURE WINDOWS BRING DAYLIGHT DEEPER INTO THE BUILDING

First Tech Federal Credit Union, Hillsboro, OR

Source: Swinerton Mass Timber

A study performed by the Energy Studies in Buildings Laboratory (ESBL) at the University of Oregon provides evidence that exposed wood supports the thermal and visual comfort of building occupants. The study found that "visually 'pleasant' or 'warm' surroundings can improve perceived thermal comfort, even when the space may call for cooling."

Researchers investigated the perception of thermal comfort in the presence of wood versus white-painted drywall in a climate-controlled chamber. After a 40-minute acclimation period in which the materials were covered with black curtains, the drywall or wood surfaces were exposed. At intervals, the test subjects answered survey questions related to comfort and perception. In the wood room, with no other variables altered, participants were 25 percent more likely to desire no change in thermal environment, or, in other words, to be comfortable.

An even stronger response was measured with a word association test. Participants related word pairs, "reveal[ing] that people found the wood walls to have more favorable qualities all-around than the white." The researchers found that "wood was considered more 'natural' than white walls or the control. Wood was also significantly more 'liked' than 'disliked' as compared to the white walls. Wood was also found to be significantly more 'expensive,' 'pleasant,' 'sturdy,' 'unique,' 'interesting,' 'new,' and 'clean' than the white."

7.2.5 VISUAL COMFORT

Key factors in the visual comfort of building occupants are visual access to nature and the amount of daylight allowed. Research shows a link between access to daylight and improvements in mood, productivity, and sleep patterns. Views can dramatically affect mood and productivity as well. A building designed to maximize daylight access for occupants will be oriented to take advantage of daily and seasonal sunlight patterns. It will also limit floor plate depth, so occupants spend most of their time near the perimeter of the building where daylight is most prevalent. Mass timber supports good design practices, with thin floor plates for higher ceilings and two-way spans that can eliminate perimeter beams. Both qualities allow for more and taller windows to allow daylight further into a building. Mass timber often inspires building designs with open atriums that are visually appealing and filled with natural light. This is related to mass timber's great strength-to-weight ratios, which enable longer spans.

7.2.6 LIFE SAFETY

Building codes ensure that occupants are as safe as possible from catastrophic events, such as earthquakes, fires, and high winds. Wood performs very well relative to building code standards, and it goes even further by contributing to highly "resilient" designs. Resilient buildings recover quickly from such disaster events. Buildings that can be safely occupied following a disaster are invaluable to recovering communities, a fact made painfully clear every time a large-scale disaster displaces a large number of people for long periods.

7.3 BEHAVIOR

When people are healthy and comfortable, they are much more likely to exhibit behaviors that benefit them and the people around them.

7.3.1 ECONOMIC BENEFITS

Terrapin Bright Green's "The Economics of Biophilia" states: "The main causes for deficient productivity include absenteeism, loss of focus, negative mood, and poor health. The built environment, though not always the cause of these stressors, when well-designed, can be a reliever of these undesirable symptoms," and that "10 percent of employee absences can be attributed to architecture with no connection to nature." Many employers understand the financial and social benefits of a healthy workplace on employee productivity and will seek spaces that best meet their needs.

Benefits are likewise present in retail environments. "Retail customers judge businesses surrounded by nature and natural features to be worthy of prices up to 25 percent higher than businesses with no access to nature." An environment where customers feel both relaxed and stimulated will be more conducive to spending, contributing to the success of a business. The ESBL study cited above in the section on thermal comfort also found that test participants perceived wood surfaces as being "expensive" and "pleasant," which also has implications for customer behavior.

Building maintenance is an expense, and occupant behavior can have a direct impact on maintenance costs. Occupants who enjoy a space and feel respectful toward a building will be less likely to be careless or destructive.

7.3.2 SOCIAL BENEFITS

The same effects that the presence of trees and green spaces have on lowering violent and criminal behavior in communities can be seen inside than in the previous building, and even more challenging to remove. Before the new building opened, a behavior strategy of quiet voices was planned for and encouraged in the halls using graphics, words, and quotes reminding students to be peaceful and wise. To the administration's delight, the students were remarkably calm and respectful in the new space. Behavior issues and subsequent disciplinary actions decreased significantly. Students report that the space makes them feel valued.

A survey in British Columbia similarly found that wood surfaces are less likely to be vandalized than other surfaces.

Though more research has been done on office environments and hospitals, focusing on productivity or infection, researchers of biophilic effects agree that the potential for schoolchildren to benefit from the healing effects of natural materials is very promising.

FIGURE 7.4: EXPOSED WOOD ENCOURAGES CALM, RESPECTFUL BEHAVIORS

William Perkin Church of England School
Source: Emily Dawson

buildings as well, reducing vandalism and other aggressive behaviors.

One mass timber example is the William Perkin Church of England High School, completed in 2014. It is constructed with exposed Cross-Laminated (CLT) walls and floors as an economic strategy to meet a tight 12-month construction schedule. The new building replaced an outgrown and dilapidated predecessor and serves a student body with noted behavior issues. There were concerns about how the new building would be treated, as vandalism may be as tempting—or even more tempting—on the new exposed wood walls

Mass Timber Innovations

From CLT, GLT, and LVL, to MPP and more

USNR is a comprehensive supplier of equipment and technology for many innovative mass timber products like Mass Plywood Panels (MPP) to large-scale CLT production, including finger-jointing, material handling, grade sorting, lay-up, and panel pressing.

USNR systems are expertly integrated to achieve an efficient and automated operation. Ask about our controlled migration program that takes you from manual CLT panel lay-up to a fully-automated panel assembly system. It's a cost-effective entry point to the global CLT market.

Our patented CLT press has a modular design that allows for infinite expansion along the length. It can grow with your operation, enabling you to meet the needs of the market today. *Contact us to learn more about our equipment and technology for mass timber.*

Millwide. Worldwide. | +1.360.225.8267 | usnr.com

USNR

CHAPTER 8: OWNERS AND DEVELOPERS

IMPACTS OF THE MASS TIMBER EFFECT

- In the near future, the carbon impact of any investment likely will factor into its market value.

- The development of forest carbon markets has the potential to inform timber use in the building industry.

- Sustainably harvested wood fits naturally into a circular carbon economy.

- Mass timber consumers who support sustainable forestry practices and policies will push the market toward the maximum carbon storage potential of forest products.

- Resilient, high-value buildings support communities facing natural disasters by providing immediate, or quickly available, safe shelter.

- At this stage in the evolution of mass timber, building owners are the true pioneers in adopting a relatively new building technology while exploring evolving financing and procurement systems. Contractors, designers, and engineers—depending on the region—may have limited experience with wood structures, though educational resources are rapidly being established nationwide. This chapter explores the owner's role, the benefits of choosing a mass timber system, key development issues, and best practices.

8.1 CARBON CONSIDERATIONS

The growing market for buildings created from sustainably harvested timber has the potential to contribute significantly to a sustainable forest economy.[1] The use of Climate Positive Forest Products (CPFP)[2] in buildings has the potential to support climate mitigation goals in three ways:

- Carbon sequestration: absorbed from the atmosphere during the growth of biomass

- Carbon storage: in bio-based materials used in long-term applications

- Carbon substitution: avoiding emissions by replacing high-embodied-carbon materials

Quantifying carbon from the standpoint of materials investment is simple in theory: if a thing takes less carbon to produce and deliver than the carbon embodied in its material makeup, it is a carbon storage device. Bio-based materials that pull carbon from the atmosphere while they grow are intriguing as carbon storage mechanisms. A bio-based product potentially has value in a market that values carbon. Buildings are massive materials repositories, driving significant investigation of the potential to quantify and capitalize buildings in the context of a carbon market.

Developers interested in this potential should become familiar with the concepts of carbon markets, carbon offsets, and a carbon economy. We explore them in the following sections.

---

1   *Built By Nature Policy Plan*, 2021-2023, The Netherlands.
2   *The Forests Dialogue, Scoping Dialogue on Climate Positive Forest Products, Co-Chairs Summary Report*, 2021.

Knowledge gaps remain on how scaling up the use of mass timber will impact forest health and carbon emissions. Awareness of these points is valuable as building owners navigate materials sourcing and public awareness of timber buildings. The Forest Dialogue[3] has identified three main areas of uncertainty:

- Whether increased demand for timber in construction is ultimately good or bad for forests

- Whether more use of wood is rarely or mostly good for the climate

- Whether to ramp up use of construction timber now or to first build evidence for positive impacts

As the industry progresses, the Forest Dialogue and many other groups are working to gain clarity around these points through research, public outreach, and collaboration.

For more definitions of many concepts around tracking and evaluating embodied carbon options, reference the section on Carbon Considerations at the beginning of Chapter 5.

8.1.1 THE CARBON ECONOMY

It is likely that, in the near future, the carbon impact of any investment will factor into its value. Carbon taxes, carbon credits, and low-carbon incentives are not yet the norm for forest products or the building industry, but they likely will be increasingly incorporated into the overall economy. In 2020, Architecture 2030 began including embodied carbon metrics in the voluntary AIA 2030 Commitment reporting program, and 291 projects in the US completed this aspect of the report.[4] According to Architecture 2030, "It's now possible for every new building to have zero-net-carbon operations. We must also dramatically reduce the embodied carbon in infrastructure, buildings, and materials—in the next 10 years."[5]

Future-minded organizations in the building industry are laying the groundwork for meaningful engagement with a carbon economy through education, tool kits, and evolving policies that support sustainable construction. Sustainably sourced mass timber buildings can potentially neutralize or even offset the carbon emissions required to construct a building—something to consider for projects that will start a permitting process in the future.

Carbon Taxes and Cap-and-Trade

The philosophy behind taxing carbon emissions is to increase the inherent value of efficient and sustainable industrial processes. Large emitters pay penalties for the carbon they use and therefore have an incentive to reduce their carbon use and the associated taxes. This approach is used primarily in Canada and Mexico.

The cap-and-trade system recognizes a market price for emissions and provides credits to companies that invest in reducing emissions but imposes a cost on remaining emissions. Companies can agree to trade credits, allowing a company that pollutes more to purchase credits from another

---

3   *The Forests Dialogue, Scoping Dialogue on Climate Positive Forest Products, Co-Chairs Summary Report, 2021.*
4   The American Institute of Architects, 2030 by the Numbers: *The 2020 Summary of the AIA 2030 Commitment* (December 2021).
5   *Architect Magazine*, The Carbon Issue, January 2020, guest edited by Architecture 2030.

that pollutes less. The lower-emitting company, therefore, has an increased inherent market value simply by being more energy-efficient.

North American cap-and-trade markets happen primarily through auctions held by trade groups: the Regional Greenhouse Gas Initiative (RGGI), which comprises 11 1Northeastern and mid-Atlantic states; and the California Air Resources Board (CARB), which trades with Québec and Ontario. Revenues are typically applied directly toward emissions-reduction projects. These entities establish a "cap," or carbon allowance, based on the size of the participating regions, and reduce the allowance over time to meet certain targets.

Carbon credit values are on the rise; RGGI reported a 23 percent rise from 2018 to 2019, and an 18 percent rise from 2019 to 2020.[6] CARB reported that the program was achieving its objectives and would become more stringent in 2021 to help the state reach its goal of reducing greenhouse gases to 40 percent below 1990 levels by 2030.[7]

Connecticut, Delaware, Maine, New Hampshire, New Jersey, New York, and Vermont jump-started the creation of RGGI back in 2005 by signing a memorandum of understanding with the intent to reduce carbon emissions. In 2007, Maryland, Massachusetts, and Rhode Island joined in as well. Virginia joined in 2021, and Pennsylvania is set to join in late 2022.

Carbon Offsets and Banking

Complementary to the cap-and-trade approach to carbon management, carbon offset programs are also growing as carbon accounting becomes more important to economies around the world. Capturing offset for a given project (e.g., forestry or a building) is possible when a given approach demonstrably goes beyond baseline industry standard performance. Forest-based carbon offsets are rapidly developing to provide landowners with an inherent value for sustainably managed landholdings. Placing a high intrinsic market value on land that might otherwise be converted to other uses is one of the many benefits of this paradigm. Finite Carbon[8] is one such program, and it reports a portfolio of 3.1 million acres and $720 million.

A forest may in fact be most valuable to a landowner as a carbon bank. Inquiries about using mass timber buildings as carbon banks are developing, though the complexities of quantifying the myriad products and material sources in a building make that approach less straightforward. Chapter 5 explores the tools available to design teams to track these metrics. The United Nations Intergovernmental Panel on Climate Change (IPCC), however, recognizes in-use Harvested Wood Products (HWPs) as one of two "non-ecosystem" forest carbon categories, the other being material in landfills, according to Dovetail Partners' 2021 report.[9] This report also notes, "Up to two-thirds of the [carbon] market could be served by natural climate solutions. However, current voluntary markets are only providing 8-13% of what the domestic market may demand within the decade and may need to grow by more than 15-fold by 2030 to support the 1.5°C pathway [to mitigate catastrophic climate change]."

6   6 Potomac Economics, *Annual Report on the Market for RGGI OC2 Allowances: 2020* (May 2021).
7   https://ww2.arb.ca.gov/resources/documents/faq-cap-and-trade-program
8   https://www.finitecarbon.com/
9   Dovetail Partners, *Carbon Storage, Credit Markets, and Forests Report* (September 2021).

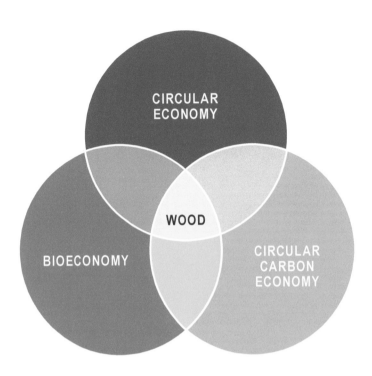

FIGURE 8.1: WOOD AT THE NEXUS OF SUSTAINABLE AND REGENERATIVE ECONOMIES

Where do forest products fit into this growth potential? Using more bio-based materials is a promising start, and it's something the building industry can undertake immediately. To truly quantify impact, we need to better understand our baselines and improve the methodologies for accounting for the carbon impacts of wood products as part of a complete, delivered building project.

A Circular Carbon Economy

The Consortium for Research on Renewable Industrial Materials (CORRIM)[10] recognizes wood as a material uniquely poised to solve the global economic, environmental, and social pressures associated with the building industry. The consortium engages researchers and practitioners to identify the carbon impact of wood products from extraction to disposal or reuse, and to propose methods to improve industry practices to maximize the "triple-bottom-line" benefits. CORRIM describes the circular economy as a way of minimizing or eliminating waste across the life cycle of a product or material. It has identified wood as central to a framework that also considers a bioeconomy (renewable biological materials) and a circular carbon economy (see **Figure 8.1**).

CORRIM notes that wood as a building material is unique because it "can be designed to be cycled through both technical and biological cycles" and because "circularity is further extended from the waste stream through the uptake of greenhouse gasses during new forest growth."

10 www.corrim.org

Quantifying Carbon in Wood Structures

Life Cycle Analysis (LCA) tools approximate the embodied carbon and carbon emissions of wood products. Because sourcing techniques vary widely, most LCA tools use aggregate assumptions that may or may not accurately reflect any one specific wood product. See Chapter 5 for more on using LCA tools for wood structures.

Rigorous whole-building or structural LCA studies are becoming more numerous as the mass timber industry matures. Collectively, these studies indicate that wood products do significantly contribute to lower embodied-carbon construction. To cite two examples, a study of Katerra's Catalyst Building[11] in Spokane, Washington, determined that the building's wood product carbon stores nearly offset the up-front carbon required to construct it. Another completed office building, Platte Fifteen in Denver, Colorado, found a 70 to 76 percent savings in Global Warming Potential (GWP) by choosing a mass timber structural system compared to primarily steel or concrete systems, and for a negligible (2 percent) cost premium.[12] Over 99 percent of the GWP in that system was contributed by the reinforced concrete topping slab.

8.2 MARKET DEVELOPMENT: US MASS TIMBER PROJECTS

The ongoing COVID-19 pandemic continues to impact construction reports worldwide. While project billing and unemployment trends indicate a recovery from last year's decreases in construction activity, high rates of inflation defined the North American construction market, with annualized rates averaging 7.42 percent—and over 13 percent in some regions.[13] Labor shortages, materials shortages, transport bottlenecks, and other supply chain disruptions are seen as significant factors in the rising costs. Globally, single-family housing starts remain extremely strong, and nearly all other building market sectors experienced a modest increase in activity from 2020. Governmental infrastructure stimulus activities are expected to contribute to the overall economic recovery in the US construction sector.[14]

Despite the ongoing uncertainties created by the pandemic and increasing climate change-related events, the number of mass timber buildings in design and construction continued to grow in 2021. The following data was provided by WoodWorks, which offers free one-on-one project assistance for nonresidential and multifamily wood buildings. Technical experts offer support, from design through construction, on issues such as allowable heights, suitable areas for different construction types, structural design, lateral systems, and fire- or acoustical-rated assemblies. WoodWorks has provided input on most of the mass timber structures designed and/or built in North America in recent years. It also tracks details related to mass timber projects.

The following figures illustrate the development of the mass timber industry in the US and provide insights on the popularity of primary materials, the regional popularity of mass timber, occu-

11 Carbon Leadership Forum (CLF) and Center for International Trade in Forest Products (CINTRAFOR), *Life Cycle Assessment of Katerra's Cross Laminated Timber (CLT) Catalyst Building* (2020).
12 https://www.klaa.com/sustainability-lcas
13 Rider Levett Bucknall, *North America Quarterly Construction Cost Report, Fourth Quarter 2021*.
14 Rider Levett Bucknall, *COVID-19 Global Sector Report, Issue 7*.

pancy types, building sizes, and the total square footage and number of projects constructed from 2012 through 2020. **Figure 8.2** illustrates the rapid growth of mass timber building projects, broken out by mass timber type. On a project count basis, most of the growth has been in the use of CLT.

Figure 8.3 shows the same information, but rather than reporting the number of buildings, this report is based on total constructed square footage. In 2021, mass timber projects totaled 5.7 million square feet. Combining data from these two figures reveals the average project in 2021 was 52,000 square feet. CLT accounts for 71 percent of the square footage and 65 percent of the building projects.

Figure 8.4 illustrates the mix of mass timber building occupancies in the US as the total constructed square footage each year (by construction start date) for each use type. Business occupancies represented 42 percent of the total built square footage in 2021, up from 32 percent in 2020. Multifamily uses dropped to 24 percent this year from 36 percent in 2020. Educational and assembly uses follow as the next-largest sectors, at 13 percent and 15 percent, respectively.

Finally, **Table 8.1** shows the number of mass timber projects in the US by state. The number of projects either under construction or completed nearly doubled in 2020; the trajectory slowed in 2021 but still saw an impressive 30 percent increase. The count of proposed projects continued to increase by almost 20 percent as well, despite the recessionary climate, indicating that growth of the mass timber market will continue for the foreseeable future.

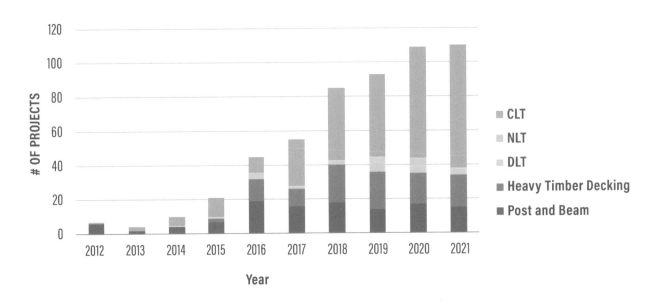

FIGURE 8.2: UNITED STATES PROJECTS BY PRIMARY MASS TIMBER MATERIAL

Data provided by WoodWorks

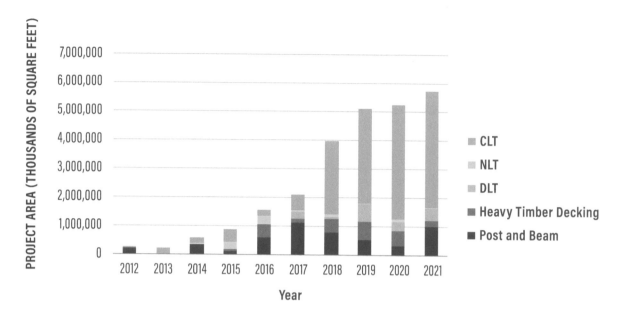

FIGURE 8.3: UNITED STATES BUILDING SQUARE FOOTAGE BY PRIMARY MASS TIMBER MATERIAL

Data provided by WoodWorks

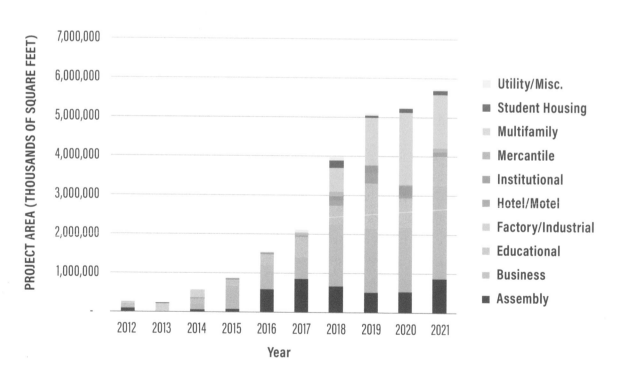

FIGURE 8.4: UNITED STATES MASS TIMBER BUILDING SQUARE FOOTAGE BY OCCUPANCY

Data provided by WoodWorks

Figures **8.5** and **8.6** combine the 10 years of data depicted in the previous charts to show total square footage and total number of projects in the US by occupancy type.

8.3 RATIONALE AND MOTIVATION

It is important to understand an owner's rationale and motivation for selecting mass timber as a building technology. In a 2014 survey[15] of tall wood building owners worldwide, the most cited motivations were market leadership and innovation, the environmental benefits associated with wood, and construction schedule savings. Owners must balance those rationales with their responsibility to seek the best return on their investment and the need to deliver a building within the allotted time frame, while ensuring the safety of construction workers and building occupants. As expertise grows in the Architecture, Engineering, and Construction (AEC) community and more mass timber projects go to market, successes are helping to allay the perceived risks.

8.3.1 BUILDING VALUE

Mass timber market data is limited by the relatively small number of buildings and the short amount of time these buildings have been on the market. However, mass timber buildings have been shown to perform well in terms of lease-up rates, tenant retention, sales, and market premiums. It is very likely that these buildings perform well because of the topics discussed in Chapter 7, the biophilic and human health benefits of being near natural materials.

15 Perkins+Will, *Survey of International Tall Wood Buildings* (2014).

STATE	STAGE		
	CONSTRUCTION STARTED / BUILT	IN DESIGN	GRAND TOTAL
AK		2	2
AL	8	12	20
AR	7	9	16
AZ	2	3	5
CA	79	120	199
CO	19	19	38
CT	9	6	15
DC	5	8	13
DE	1		1
FL	23	39	62
GA	13	19	32
HI	2	1	3
IA	5	6	11
ID	7	3	10
IL	15	19	34
IN	3	1	4
KS	2	2	4
KY	5	1	6
LA	1	5	6
MA	25	40	65
MD	5	10	15
ME	6	9	15
MI	2	18	20
MN	11	3	14
MO	9	9	18
MS	1	3	4
MT	14	9	23
NC	31	34	65
ND		1	1
NE	3	11	14
NH	1	3	4
NJ	3	8	11
NM	1	1	2
NV		2	2
NY	15	36	51
OH	9	10	19
OK	3	2	5
OR	66	28	94
PA	7	11	18
RI	3	2	5
SC	19	8	27
SD	1	1	2
TN	9	6	15
TX	36	55	91
UT	8	11	19
VA	8	13	21
VT	2	10	12
WA	73	53	126
WI	21	18	39
WV	2		2
WY	3		3
Grand Total	**603**	**700**	**1303**

TABLE 8.1: US MASS TIMBER PROJECTS BY STATE
Data provided by WoodWorks

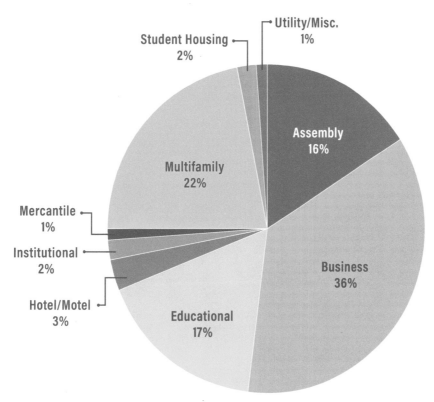

FIGURE 8.5: US TOTAL MASS TIMBER BUILDING
SQUARE FOOTAGE BY OCCUPANCY TYPE

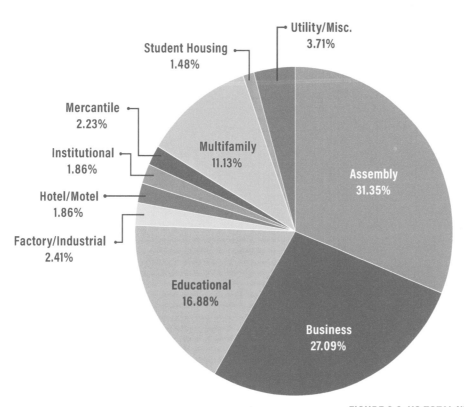

FIGURE 8.6: US TOTAL NUMBER OF MASS TIMBER
BUILDINGS BY OCCUPANCY TYPE

Source: PAE; Photo Credit: Jamie Goodwick/Portlandrone

CASE STUDY: PAE LIVING BUILDING

BUILT TO LAST 500 YEARS

STORY CREDIT: NADINE M. POST OF ENGINEERING NEWS-RECORD AND PAE

LOCATION: PORTLAND, OREGON

COMPLETION DATE: FALL 2021

OWNER: PAE, EDLEN & CO., ZGF ARCHITECTS, AND WALSH CONSTRUCTION

DEVELOPER: EDLEN & CO.

ARCHITECT: ZGF ARCHITECTS

STRUCTURAL ENGINEER: KPFF

CONTRACTOR: WALSH CONSTRUCTION

MT MANUFACTURER(S): STRUCTURLAM MASS TIMBER

MT INSTALLATION: CARPENTRY PLUS

CERTIFICATIONS: PURSUING LIVING BUILDING, PURSUING ARCHITECTURE 2030, PURSUING NET ZERO WATER, PURSUING NET ZERO ENERGY

THE FIVE-STORY PAE LIVING Building, named after its prime tenant, is on course to be the largest privately owned office building to be fully certified under the Living Building Challenge (LBC), the world's most stringent sustainability standard. It is

Source: PAE; Photo Credit Credit: Oregon Department of Forestry's upcoming "The Promise of Mass Timber"

also the first developer-led commercial Living Building, proving the business case and empowering others to follow a similar path.

This urban office development is one of a very few spec office buildings and the only Living Building engineered beyond code minimums for hospital- and fire station-level seismic resilience.

Portland is in earthquake territory. The structure, designed to last 500 years, followed the prescriptive code for seismic Risk Category (RC) IV rather than RC II, the minimum required for office buildings. The 5-ply Cross-Laminated Timber (CLT) floor slabs are topped by a 3.5-inch-thick slab that transfers lateral loads to the concrete shear core and doubles as a radiant floor for heating and cooling. Glulam beams span 32 feet and are supported on interior glulam columns and perimeter columns.

In the summer months, the building will be able to function for 100 days without using any city utilities,

thanks to a large rooftop solar array and battery storage. In addition, an on-site waste recovery system turns liquid waste into fertilizer and solid waste into compost. The building could also act as an emergency operations center in the event of a major quake.

The durable and beautiful timber structure reduces the project's embodied emissions by 30 percent and contributes meaningfully to the strategy of 500-year building design. The modular timber system is easy to maintain, and the building organization and interior design offer significant flexibility for multiple use types and tenants in the future. The project team understands that buildings are preserved because they are loved, and some of the oldest buildings in the world still standing today are built from wood. The timeless design and brilliant utility of the PAE Living Building ensures it will be cherished and preserved for many generations to come. ⬤

Buildings of the Future

Environmental and carbon sequestration credentials will be attractive to a growing market of environmentally conscious tenants and buyers, particularly in the home and corporate markets (see **Figure 8.7**). These buildings may also have a place in the carbon markets discussed earlier in this chapter.

Leasing and Tenant Retention

Because of the increased demand for biophilic buildings, the leasing period for exposed mass timber buildings (**Figure 8.8**) can be lower than for a typical concrete or steel building with traditional finishes. Securing tenants early allows the building to more quickly reach stabilization, when the building is at full occupancy and generating regular income. After stabilization, the loan payment (including the interest) is covered by the income, allowing a building owner and/or investor to begin recouping their investment. Once the building is stabilized, permanent financing can be obtained at a fixed interest rate or the building can be sold. The earlier the building is fully leased, the better the return on investment.

TOP — FIGURE 8.7: LARGE CORPORATE BUILDING OWNERS ARE TURNING TO MASS TIMBER

Source: Microsoft, Holmes structures;
Photo Credit: Blake Marvin Photography

BOTTOM — FIGURE 8.8: RADIATOR BUILDING

Source: Andrew Pogue Photography

FIGURE 8.9: ONE NORTH DEVELOPMENT

Source: Kaiser + Path

Studies have not yet been done to provide aggregated data on the market success of mass timber real estate, but as the market matures, the anecdotal evidence is increasingly positive. The Platte Fifteen LCA study mentioned earlier in this chapter also reflected the economic success of the project; the building was 85 percent leased only 1 month after completion, and at higher lease rates than any comparable building in the area. The Portland, Oregon, area has the highest concentration of mass timber building projects in the US, and local developers are reporting significant leasing advantages for these office buildings:

- Beam Development opened their District Office in 2020 and was 66 percent leased at project completion, double comparison expectations of 20 to 30 percent. As of January 2022, the building was 73 percent leased. 811 Stark opened in 2015 with 100 percent of the office space leased, and the retail spaces leased shortly thereafter. The building has remained completely leased since opening.[16]

- PAE Living Building, developed by PAE and Edlen & Co., opened in fall 2021 with 88 percent of the office space and 25 percent of the retail space leased.[17]

- Killian Pacific's growing mass timber portfolio includes Nova and The Hudson, both opened in 2016 and stabilized in 2 and 5 months, respectively. Two years later, Convene was completed and stabilized at opening. Skylight was also completed in 2018, quickly leased to 84 percent, and stabilized at 14 months.[18]

- One North, a collaborative development by Karuna Properties II, LLC and Kaiser Group, Inc. (see **Figure 8.9**) was completed in 2014, with key anchor tenants committed before groundbreaking. Even with unprecedented lease rates for the east side of Portland and very little parking, the buildings were fully leased 6 months faster than the pro forma had assumed.[19]

In addition to faster lease-up rates, mass timber buildings can demand premium rental income. Exposed wood ceilings are a premium finish when compared with painted drywall or concrete. Floor-to-ceiling dimensions can be greater due to the strength and spanning capacity of the panels and the beauty of exposing the structural deck. These factors contribute to higher lease rates for little to no added construction cost, translating to a higher sale price for the building in the long term.

16 Information courtesy of Beam Development.
17 Information courtesy of PAE and Edlen & Co.
18 Information courtesy of Killian Pacific.
19 Information courtesy of Kaiser + Path.

Premiums

When there is a comparative cost increase associated with using mass timber over other structural systems, the premium should be balanced by adjusting the pro forma to include increased market value, illuminating payback periods. The Canyons, a 6-story apartment building completed in late 2020 in Portland, Oregon, compared a CLT structure to light framing and painted drywall. The team discovered that the payback period for the premium structure was just over 3 years, and the project proceeded with the mass timber option. Ensuring premium market differentiation with a short payback period justified the relatively small capital cost increase.

End-of-Life Value

A building that consists of high-quality modular components that can be easily reappropriated for new uses will have an inherently higher value at the end of its life than a building slated to go entirely to the landfill at demolition. Design for Disassembly (DfD) is a growing area of understanding for designers and builders, and one that a building owner may be inclined to pursue as a point of interest for future buyers.

Though it is far too early to have data on the deconstruction advantages of the recent wave of mass timber construction, reuse potential is likely to be a unique asset as these buildings age. Most other primary structural systems are difficult and costly to salvage, and often total demolition is the only viable solution from a cost standpoint. When salvage is possible, material is not usually reused as a complete element but as recycled ma-

terial in newly formed components. But, like large steel members, salvaged and reused mass timber elements could have viable market use with much less reconfiguration.

There are a couple of important issues to resolve before mass timber panel buildings reach their end-of-life in substantial numbers[20]:

- Current practice promotes long, self-tapping screw connectors that are strong and easy to install. However, they are difficult to remove without damaging the panel perimeter.

- Current practice favors concrete finishing of hybrid mass timber panels in certain classes of public and industrial buildings. It is also common in residential buildings for vibration and impact sound mitigation. These integrated floors may pose difficulties to orderly disassembly and may preclude reuse.

- All mass timber panels are custom-produced for specific products. No market exists for the production of blank panels unassigned to specific projects. It is reasonable to presume there will be some difficulty in finding a market for panels prefabricated for a decommissioned project, especially those that are integrated with multiple other materials and trades, such as windows, doors, conduit openings, and connector nests.

These concerns are all possible to address through a DfD process, promoting circular use of decommissioned elements. DfD is achieved through mindful design and detailing. Investors can lead the process by emphasizing the importance of designing pathways for cost-effective deconstruction and reuse

20 L. Muszyński, M. Riggio, M. Puettmann, A. Dodoo, L. Schimleck, and N. Ahn, *Conceptualizing the End of Life for Mass Timber Panel Buildings towards Circularity: Mapping the Gaps in Knowledge* (2021).

potential of recovered elements. An architectural design team following DfD principles may consider the potential for reusing entire subassemblies to reduce the substantial costs of refabrication and related waste. Cascading use of recovered elements should be planned during the initial design of the building, well ahead of deconstruction.

A building that has been designed for disassembly will also be easier to retrofit or repair after a disaster. For more on this topic, see the section on resiliency later in this chapter.

8.3.2 INCENTIVES

Incentives for sustainable and low-carbon buildings vary by jurisdiction and project type. Choosing mass timber construction may have associated financing or zoning incentives (such as increased floor area ratio [FAR]) for reduced embodied carbon or innovative technologies.

8.3.3 MAXIMIZE ALLOWABLE BUILDING AREA

Mass timber structures create opportunities within established zoning constraints as well. A timber building on average weighs only 20 percent as much as a steel or concrete structure. On sites with challenging soil conditions and bearing pressure limitations, a lighter building could be built larger or taller than a heavier building. The lighter weight can be particularly advantageous in high-seismic-activity regions. In areas where foundations to support a heavier building are prohibitively expensive, a lighter building may be the difference that makes a project viable.

Another opportunity for overall building area increases is added floors because of reduced floor-to-floor heights. Mass timber floor sections can be designed more thinly than other options and have inherent fire resistance, requiring no added fireproofing layers at certain building heights.

8.3.4 TALL TIMBER AND COST-EFFECTIVENESS

Because light framing is competitive for many low-rise buildings, and mass timber is consistently cited as competitive with concrete under 20 stories, a so-called "sweet spot" has emerged for mass timber somewhere between 4 and 18 stories, depending on the market in question. With increasing urban density, the largest market growth for new buildings in the coming years is projected to be in the mid-rise range, between about 3 and 8 stories. WoodWorks reports that, as of January 2022, over 120 mass timber projects are actively in design in the US, and over 50 percent are multifamily projects of 6 to 12 stories. Mass timber is poised to be a competitive option for a majority of foreseeable increases in building stock.

While mid-rise construction will continue to be the most common new building stock for all construction types, buildings over 20 stories are impactful from both a market and an environmental resource standpoint. Using mass timber for buildings above the current US code limit of 18 stories has increasing potential. The tallest mass timber buildings in the world use CLT and glulam as the primary structural materials and concrete for cores and/or additional mass:

- 18 stories, 174 feet (53 meters): Brock Commons, University of British Columbia, Vancouver, BC (2017)

- 24 stories, 276 feet (84 meters): HoHo Vienna, Woschitz Group, Vienna, Austria (2019)

- 18 stories, 279 feet (85 meters): Mjøstårnet, AB Invest, Brumunddal, Norway (2019)

- 25 stories, 284 feet (87 meters): Ascent, New Land Enterprises, Milwaukee, Wisconsin (2021)

Additionally, a growing number of studies and proposals are validating the effectiveness of timber structures up to 35, 40, and even 80 stories.[21,22,23]

Allowable timber building heights increased in the 2021 International Building Code to 9, 12, and 18 stories, with varying amounts of exposed wood allowed. The next round of updates, to be formalized in 2024, increases the allowance of exposed wood surfaces (see Chapter 5 for more information). Well-designed wood buildings that are taller or include more exposed wood than codes allow can be proven viable and safe. Depending on the jurisdiction, such designs may be permissible through an alternate means and methods, performance-based permitting approach.

The Council on Tall Buildings and Urban Habitat (CTBUH) is developing resources for project teams pursuing tall mass timber buildings, supported by grant funding from the US Forest Service. The group worked to establish the inclusion of timber projects within the CTBUH height criteria and published *Timber Rising*, combining the best research and resources for tall timber projects.

8.3.5 CONSTRUCTION RISK REDUCTION

The modularity, precision, and beauty of large, engineered timber components has refreshed conversations about the benefits of off-site construction for others. When a modular structural system like CLT is assembled in half the time of a traditional structure with a lower risk and a higher level of craftsmanship, designers and builders start to look for ways to shift the fabrication of other components into more controlled environments.

Site-built construction is often challenged by weather, traffic, noise ordinances, labor shortages, and any number of physical site constraints, as in **Figure 8.10**. Customized prefabrication can alleviate these issues, depending on the project and the extent to which the design and build team can plan and coordinate it. The resulting building can have a higher level of precision than site-built structures because of the increased quality control afforded by climate-controlled factory environments.

Chapters 5 and 6 go into depth on the advantages of off-site fabrication and the design processes and collaboration necessary to achieve success. In short, taking more time up front in the design phase pays off in construction-phase speed and predictability. The precision of custom components and a highly organized modular structural package contribute to expedited construction with fewer field modifications, change orders, and delays.

Considering that a building's superstructure is usually about 20 to 25 percent of the total construction cost, investing in a highly predictable assembled structure may significantly reduce risk. Mechanical, electrical, plumbing, and fire systems account for another 30 to 35 percent of a building's cost, or about 15 percent for core-and-shell projects. These systems may or may not

21 https://www.gensler.com/blog/developing-worlds-tallest-net-zero-timber-building-sidewalk
22 https://perkinswill.com/project/canadas-earth-tower/
23 https://perkinswill.com/project/river-beech-tower/

FIGURE 8.10: SIDEYARD

Source: Project: Sideyard; Photo Credit: Skylab Architecture

also be fabricated off-site for schedule savings. If well-coordinated with the structure, the associated change risk also goes down. Change cost contingencies could be reduced by up to 50 percent.

Other associated benefits include fewer potential weather delays and lower costs associated with traffic disturbances.

Carrying Costs

The construction cost savings of a modular approach, such as CLT, will be multiplied if financing impacts are considered, in addition to construction overhead and other capital savings.

Comparative information about the construction duration of different structural options can have a significant impact when applied to carrying costs such as loan interest payments, property taxes, and other fees. Reducing carrying costs by even a month or two translates to tangible savings that should be included in comparative cost models.

8.3.6 RESILIENCY

Resiliency is a term used to describe a building's ability to recover from a disaster such as an earthquake, fire, hurricane, or flood. Mass timber has several resiliency advantages over steel, concrete, and light frame structures.

Figure 8.11: CLT Rocking Shear Wall

Wall has steel fuses for dissipating seismic forces. Broken fuses are easily replaced.
Source: Project: Oregon State University Peavy Hall Replacement; Photo Credit: Andersen Construction

FIGURE 8.12: ROCKING SHEAR WALL FUSE
Source: Project: Oregon State University Peavy Hall Replacement; Photo Credit: Hannah O'Leary

Mass timber is both strong and flexible and therefore well-suited to resisting large forces and returning to its original shape. It is also very fire-resistant because of the thickness of each member. Unlike steel and concrete, failures or compromises in wood structural members are visible, so they require no special forensic equipment or destructive means for analysis, such as radar or core drilling. Being able to quickly verify the safety of a building after an event hastens reoccupancy.

Provided a design makes retrofitting damaged elements possible (see the earlier section on end-of-life value), mass timber components that show signs of compromise can be more easily replaced. Instead of an entire building being condemned, areas requiring repair can be isolated and retrofitted.

An innovative, earthquake-resistant "rocking" shear wall design has been installed in the new George W. Peavy Forest Science Center building on the Oregon State University campus, the first example of such construction in North America (see **Figure 8.11**). The design allows the wall to shift and return to place during a seismic event, with the added flexibility of steel tension rods that run the height of the wall and energy-dissipating steel "fuses" (**Figure 8.12**) that connect panels. The easily replaceable fuses are designed to break under high force, rather than allowing the destructive forces to transfer to the structure. The fuses can be easily accessed and are low-cost to replace, if necessary. Seismic damage is confined to these components.

8.3.7 MAINTENANCE AND BUILDING MANAGEMENT

Operational ease and savings can be planned for more easily when executing a modular mass timber building because of a more collaborative design phase and few changes during the construction phase. Although timber has material-specific upkeep, such as coatings, the natural beauty of wood offers some surprising benefits.

Utilities

Exposing wood is often a primary reason to use timber as a structural material. This decision should be paired with a deliberate approach to locating utilities, whether visible or concealed within chases and soffits. Mass timber buildings require more planning in the design phase, often leading to predetermined slab and wall penetrations for ductwork, conduits, and piping. This provides an opportunity to design utility systems in a building with ingenuity and precision, and it ensures that systems are installed according to plan. Having reliable as-built documents can lead to more efficient routine maintenance, and when systems issues arise, to more timely action.

Durability

Coatings such as sealers or paints may be added to structural timber to protect it from ultraviolet light and weather, to add aesthetic appeal, or to make cleaning easier. Coatings on any surface require some upkeep and reapplication. Maintenance timelines vary by product, application method, and exposure; the more the wood is protected from weather, the longer the coatings will last.

Wood naturally changes color over time, with the hue depending on exposure and species. In Europe, it is more common to let exterior wood naturally age with weather and sunlight, creating a beautiful, varied texture on a building's facade. In the US, it is more common to seek a controlled, even look. The preference is cultural, as wood that is given sufficient protection through good architectural detailing will take a long time to degrade, even without protective coatings.

Because wood is porous, many building owners are concerned about occupant damage such as staining, impact damage, or vandalism. But owners of wood buildings have reported higher levels of occupant care with wood surfaces and reduced occurrences of vandalism. (See Chapter 7's section on occupant behavior for more information.) Staining can often be easily sanded away. Depending on the species, wood surfaces may be more or less susceptible to visible damage from minor impacts. Some variation and patina will occur over time, and again it is a matter of preference whether the change is considered negative or positive.

8.4 EXECUTING AN INNOVATIVE PROJECT

While mass timber uptake in North America continues at an exceptional rate, for most markets on the continent, it is still an emerging technology. Finding an experienced team is one effective way to mitigate risks associated with innovative approaches, but strong goals and leadership on the ownership side are just as crucial. This section identifies key issues that building owners/developers face when using mass timber.

8.4.1 CHOOSING A TEAM

Investors in mass timber buildings can benefit from the early recognition that a high level of design and build team integration is a necessity, not an option. Some high points are listed below, but it is helpful to refer to Sections 5.3 (design) and 6.3 (build) for more detail on the paradigm-changing integrated approach to panelized mass timber buildings.

The British Columbia Construction Association sponsored a study of innovative technologies and strategies in building construction procurement.[24] Qualities of successful projects include:

- A highly effective and collaborative project team that puts the interests of the project first.

- Multiproject engagements of consultants and contractors to foster collaboration, learning, and team cohesion.

- Greater collaboration, leading to more successful outcomes and higher-level team performance.

- Starting the procurement process as early as possible to allow collaboration to start and creative ideas to blossom.

- Allowing the project team input on when research and development, tours, and project documentation activities can best occur from the perspective of maintaining an efficient and safe site.

- Construction Management at Risk (CRM) or Single-Purpose Entity (SPE) for Integrated Project Delivery (IPD) contracts (such as multiparty agreements) that encourage col-

laboration may be best suited for innovative projects that are not well-defined in scope.

- Requiring evidence of the qualification of individuals as part of the evaluation. The names of key project team members (including important trade companies) need to be written into the contract documents to ensure their expertise is being applied to the project and not passed on to others in their company.

- The owner ensuring it has the capacity to carry out project leadership and oversight effectively, potentially through an external project manager. Operations and maintenance personnel should also be involved.

- Encouraging businesses of all sizes to participate because some small- to medium-size enterprises are the most innovative.

- Reducing barriers to participation by simplifying the procurement process as much as possible, e.g. admitting bidders who may not have directly relevant project experience but have transferable expertise with a similar project type. Focus is on the quality of the references rather than quantity.

In summary, highly collaborative, nimble teams who are eager to innovate and willing to solve problems are more likely to achieve success with new approaches.

8.4.2 DESIGN-PHASE-FORWARD PLANNING

Mass timber is a catalyst for unusual design-phase-forward planning that can have significant impacts on construction schedules. An expe-

---

24 British Columbia Construction Association, *Procuring Innovation in Construction: A Review of Models, Processes, and Practices* (2016).

rienced team will plan for adequate coordination time before construction starts to reduce costly field labor and project overhead. Advantages to investing in early coordination include the following:

- Precision placement of mechanical, electrical, and plumbing (MEP) penetrations results in fewer trade conflicts on-site, and allows for off-site fabrication of components for rapid sequencing.

- A custom mass timber package is predictable to install and precise to a ⅛-inch tolerance. If the package is fully coordinated, it should require no field modifications.

- Change orders associated with the structure and MEP trades are minimized by up-front coordination.

Understanding the schedule savings and the reduction of on-site risk is crucial for producing an accurate cost model. According to Swinerton, "A large-scale mass timber project can be up to 2 percent higher in direct costs, but a minimum of 20 percent lower in project overhead costs. The net result is cost-neutrality and higher value."[25]

It is advisable to invest more time in the design phase to facilitate more efficient manufacturing and fabrication, to reduce construction time, and to increase construction predictability. This may have implications for how the project is financed, increasing up-front soft costs but decreasing hard costs and interest payments in construction.

8.4.3 PROCUREMENT PROCESSES

Standard procurement processes can be a barrier to maximizing the cost benefits of mass timber, discussed at length in Chapter 5 and 6.

A traditional Design-Bid-Build (DBB) procurement process is common and, as such, is preferred by many investors. For the purposes of this section, the issues are like those of a Construction Manager/General Contractor (CM/GC) process:

- Design a building to a given program, budget, and the local jurisdiction's requirements.

- Request bids from building contractors who seek best value from a variety of installers and manufacturers.

- Select a contractor (or subcontractors) to construct the building based on the apparent best value.

An effective mass timber design, however, requires extensive coordination with a procurement and installation team before putting the project out for bid. Efficiencies in materials layout and site logistics can be incorporated into early cost estimates accurately only if an experienced team is consulted. It is possible to design a mass timber building with average assumptions about efficient fiber use, fire ratings, cost, and availability. However, this approach carries risks because of possible delays and costs associated with redesign further along in the process, including design fees, permit revisions, constructability issues, and materials availability. The earlier a procurement and installation team is brought on board, the

---

25 Erica Spiritos and Chris Evans, Swinerton Builders, "Mass Timber Conference 2019 Presentation: Mass Timber Construction Management: Economics & Risk Mitigation."

more refined and cost-effective the design and construction process will be.

One option in a traditional DBB contract model is to partner with a manufacturer during the design phase using a separate contract or a letter of intent to select that manufacturer during bidding. This can be done as an agreement with the owner or with the CM/GC. The advantages of this approach include design optimization, detailed pricing feedback during design, and early assurance of product delivery dates. The risks include lack of precedent, resulting in limited manufacturer availability during design for fabrication teams who are unaccustomed to design team integration. But remaining flexible until a project is ready to order can have advantages in a changing market. Until manufacturing supply catches up with the increasing demand for mass timber products, the lead time for detailing on the manufacturer's end can be a deciding factor.

Building owners may also choose a different, more inherently collaborative procurement model to avoid these issues and support an integrated design process. For example, Design-Build, where the contractor and the design team are chosen and contracted together, or Integrated Project Delivery (IPD), where all parties are incentivized for project success, will naturally support early and efficient coordination. Having a design optimized early on will help ensure that fabrication timelines will be met if market demand is high. An experienced procurement team will be able to navigate these challenges.

The necessary prefabrication of massive panel elements creates an incentive for panel manufacturers to integrate along the traditional building project supply/value chain and to offer an integrated solution package rather than fabricated elements alone. As a result, many companies incorporate internal design, project management, and construction teams, or they ally with experienced companies. When possible, it makes sense for investors to consider such an integrated package, and to make sure there are good reasons for seeking alternatives.

8.4.4 INSURANCE

Insurance coverage for building owners is classified by susceptibility to damage by fire as determined by past incidence rates. Without a breadth of experience or data on mass timber buildings, the insurance industry perceives all wood buildings similarly. A lack of data, to insurance underwriters, indicates high risk. To date, mass timber structures have been grouped with light frame structures, despite markedly different performance regarding fire, seismic, and water damage. As a result, premiums are just as high as combustible construction types, though timber structures are more analogous to noncombustible types. According to a Perkins+Will study,[26] mass timber has yet to be fully recognized by the insurance industry as comparable to a concrete-and-steel structure. Efforts are underway in the insurance industry to recognize mass timber as a unique structural building category. In a 2021 paper exploring these issues, WoodWorks stated that an entirely new classification code is a possibility.[27] Other benefits that could reduce perceived

---

26 Perkins+Will, *Mass Timber Influencers: Understanding Mass Timber Perceptions among Key Industry Influencers* (October 2018).

27 WoodWorks, *Insurance for Mass Timber Construction: Assessing Risk and Providing Answers* (2021).

risks include the resiliency (the ability to swiftly recover from catastrophic events) of some mass timber designs.

Further development of moisture protection and construction schedule reference data would likely also support lower builder's risk insurance premiums. Understanding how moisture control methods are implemented and monitored could reassure providers about the level of risk involved in the construction of timber buildings in wet climates. The significant risk advantages of dramatically reduced construction times for modular structural approaches should also be a factor.

As developers turn to more sustainable portfolios, insurance offerings will naturally become more competitive. Some North American insurance companies have recognized the market and the opportunity to align with sustainable practices. Perhaps not surprisingly, European-based companies are more comfortable with the construction type, as they've had more time to build mass timber structures into their portfolios and observe how they perform. Swiss-based insurance company Zurich North America recognizes the "increasingly popular segment of the construction market"[28] and has increased builder's risk capacity in the mass timber sector.

8.4.5 COST UNCERTAINTY

The cost uncertainty associated with mass timber building projects is attributable to a combination of factors stemming from limited experience all along the supply chain. As the industry evolves, there is growing evidence that, although the materials cost for a mass timber building may be higher than concrete or steel, mass timber construction remains competitive because of labor savings, less costly foundations, reduced project and financing timelines, and more quickly realized revenue from a completed building.

The marketplace for mass timber products is increasingly competitive as the number of manufacturers grows, both in North America and abroad. The learning curve to construct with timber is relatively easy to overcome, but inexperienced builders will have difficulty estimating the savings associated with using mass timber and with learning to be a part of an up-front planning process. The number of manufacturers, designers, and builders who understand how to deliver efficient, cost-effective mass timber buildings is growing because the value of completed buildings is being proven in the marketplace.

8.4.6 PUBLIC PERCEPTION OF MASS TIMBER

According to a 2015 public survey[29] by Perkins+Will, the public perceives these factors to be the greatest barriers to wider adoption of mass timber:

- The flammability of wood

- Wood's strength compared to concrete and steel

- Deforestation concerns

The same study found that these barriers diminish as the public gains knowledge about and experience with mass timber buildings. Nevertheless,

28 https://insights.zurichna.com/
29 "Tall Wood Survey," *Perkins+Will Research Journal*, 8, no. 01 (2016).

these perceptions are often an obstacle that building developers must address.

8.4.7 SOURCES OF RELIABLE INFORMATION

While WoodWorks and other organizations have provided extensive support to mass timber building projects, a lack of reliable information about mass timber is still cited as a barrier to wider adoption of this technology. Resources in the form of handbooks, standards, best practices, case studies, and more are growing exponentially with the expansion of the market.

OSU Cascades- Edward J. Ray Hall
Dean Guernsey

Side Curve
Weyerhaeuser

Tsawwassen Shores
Harp Specialty Lumber

PDX Next
Brewington Photography

Source: Bouten Construction

CASE STUDY: VAAGEN

WHY MASS TIMBER: THREE PERSPECTIVES

LOCATION: COLVILLE, WASHINGTON

COMPLETION DATE: FALL 2020

OWNER: DELTA DENTAL OF WASHINGTON

ARCHITECT: HDG ARCHITECTURE OF SPOKANE, WASHINGTON

STRUCTURAL ENGINEER: DCI ENGINEERS

FIRE SAFETY AND CODE: BOUTEN CONSTRUCTION

GENERAL CONTRACTOR: BOUTEN CONSTRUCTION

MASS TIMBER CONTRACTOR: FOUST FABRICATION

MASS TIMBER MANUFACTURER: VAAGEN TIMBERS

Delta Dental of Washington, their architect, and their builder discuss with Vaagen Timbers why they chose to use mass timber in their new Colville, Washington, campus.

Why were you initially interested in mass timber for this project?

Delta Dental of Washington: We were looking for a product produced locally in Stevens County that would make the building stand out and make truly a "Colville" building. The architectural design of each of our campuses reflects something distinct about the area the office is located in.

HDG Architecture: The opportunity to utilize mass timber for this project excited our team at HDG for a variety of reasons. Architecturally, we were excited about the sustainability aspects inherent to the material itself, as well as the quality and longevity of the material as a structural component. From the beginning, when we started talking about CLT as a strong preference for the roof structure, we

knew that it absolutely needed to express itself as a design component and be displayed as an exposed finish. Truth in materials and the ability to create an honest building, or a building that doesn't need to hide its components beneath layers of additional costly finish, was (and is) an exciting idea for us.

Why did you ultimately choose to go with mass timber from Vaagen Timbers for this project instead of the alternative options?

Delta Dental of Washington: Vaagen Timbers' sustainable methods of production line up with Delta Dental of Washington's sustainability community.

HDG Architecture: Knowing that we could receive a "ready to install" prefinished product from Vaagen Timbers, shipped to site CNC'd to strict tolerances, meant that the project could be brought to market quite a bit more efficiently.

Bouten Construction: The owner had a clear understanding of their project budget and, as part of our target value delivery process, we were able to provide some cost certainty around what the wood frame was going to cost. Our research showed that although there was a cost premium for the preferred Vaagen

CLT option as compared to the car deck options, the team felt the benefits were worth the premium. As a project team, we felt we could find savings elsewhere in the budget to offset this premium. Vaagen Timbers is located in the same community, with its manufacturing facility near the project. Bouten works hard to utilize local partners to build iconic and important buildings for the community. In this project's case, both Delta Dental and the owner's rep were aware of Vaagen and were interested in utilizing them in the project, if it made sense to do so.

What were the results with mass timber for this project?

Delta Dental of Washington: The building looks absolutely phenomenal. The warm wood tones on all the exposed CLT and GLB just scream Pacific Northwest. The structure went up incredibly fast.

HDG Architecture: We couldn't be happier with the results of utilizing mass timber for the project. The ability to showcase the history and traditions of Colville in such a beautiful, lasting, and inherently sustainable product from Vaagen Timbers is something we remain truly proud of. ◓

MEET THE AUTHORS

ROY ANDERSON

VICE PRESIDENT
THE BECK GROUP

Roy has more than 30 years of forestry and forest products industry experience, including completion of two CLT manufacturing feasibility studies. He also spoke about mass timber lumber supply at the 2016 International Mass Timber Conference.

EMILY DAWSON

PARTNER,
PRINCIPAL ARCHITECT
KAISER + PATH

Emily has investigated and implemented mass timber solutions for nearly a decade, including designing the first Cross-Laminated Timber structure built in Oregon. At Kaiser + Path, she works exclusively on mass timber projects, from feasibility through construction

DAVE ATKINS

SUSTAINABILIST,
FORESTER,
ECOLOGIST, WRITER

Dave introduced the US Forest Service leadership to CLT in 2010 and helped launch their participation in its development. He retired from the USFS in 2014 but continues to work at the nexus of forest management and wood products to help create the balance between social, economic and environmental needs of a carbon neutral society, from biochar to mass timber to resilient sustainable forests.

LECH MUSZYNSKI

PROFESSOR
OREGON STATE
UNIVERSITY

Dr. Lech Muszyński is a professor in the Department of Wood Science and Engineering at the Oregon State University. A native of Poland, he received his MS in Wood Technology and PhD in Forestry and Wood Technology from the University of Life Sciences in Poznań, Poland. Lech joined OSU in 2004. Since 2010, one of the focus areas of his research has been the Cross-Laminated Timber (CLT) technology and other Mass Timber Panel (MTP) products. Lech has toured MTP manufacturing plants, construction sites, MTP-focused research centers, and related businesses across the globe.

PRODUCED BY

CRAIG RAWLINGS

FOREST BUSINESS NETWORK

BRYAN BECK

THE BECK GROUP

BEN KAISER

KAISER+PATH

FOR QUESTIONS ABOUT THIS REPORT OR TO OBTAIN ADDITIONAL COPIES. REACH US AT **MASSTIMBERREPORT.COM**

CASE STUDY INDEX

CASE STUDY: **FRERES LUMBER** _____ 40

CASE STUDY: **TIMBER AGE SYSTEMS** _____ 74

CASE STUDY: **EUCLID TIMBER FRAMES** _____ 84

CASE STUDY: **PORTLAND INTERNATIONAL AIRPORT MAIN TERMINAL EXPANSION** _____ 96

CASE STUDY: **PMX 15** _____ 110

CASE STUDY: **HESS TIMBER/GOOGLE** _____ 127

CASE STUDY: **BRITISH COLUMBIA** _____ 137

CASE STUDY: **IDAHO CENTRAL CREDIT UNION ARENA** _____ 156

CASE STUDY: **VAPROSHIELD** _____ 162

CASE STUDY: **SOPHIE RADICH SCHOOL** _____ 166

CASE STUDY: **MONTANA FWP CONFERENCE CENTER ADDITION** _____ 170

CASE STUDY: **PAE LIVING BUILDING** _____ 192

CASE STUDY: **VAAGEN** _____ 208

ADVERTISERS INDEX

ARXADA _____ 136

BATHSYSTEM _____ 87

BECK AMERICA _____ 206

CLEMSON _____ 31

CREE _____ 83

DCI ENGINEERS _____ XI

DCI ENGINEERS: SPONSOR SPOTLIGHT _____ VII

FAY JONES SCHOOL OF ARCHITECTURE AND
DESIGN _____ 38

FOREST BUSINESS NETWORK _____ II

FOREST SERVICE _____ 145

FRERES _____ 66

KAISER+PATH _____ 175

KALESNIKOFF _____ 73

LOCTILE _____ 213

MASS TIMBER STRATEGY _____ 172

MAXXON _____ 25

MINDA _____ XXXVI

NORDIC STRUCTURES _____ XXV

SANSIN _____ V

SFS GLOBAL _____ 147

SIMPSON STRONG TIE _____ 99

STERLING _____ I

STILES _____ XXXVII

SUSTAINABLE FORESTRY INITIATIVE _____ 46

TALLWOOD DESIGN INSTITUTE _____ 39

THE BECK GROUP _____ 60

TIMBER FRAME CONNECTION _____ 11

UNION PACIFIC _____ 13

USNR _____ 182

VAPROSHIELD _____ 161

WEYERHAEUSER _____ 65

WHOLE TREES / PORT BLAKELY _____ X

WHOLE TREES / PORT BLAKELY: SPONSOR
SPOTLIGHT _____ IX

WOODWORKS _____ 15

ZIP-O-LAMINATORS _____ 207

LOCTITE®

CLT | GLULAM | HRA AND STRUCTURAL FINGER JOINT | I-JOIST | FACE GLUING

One-component adhesive makes handling & processing efficient

Meets all fire performance requirements

Performs well in different environmental conditions

Product certified for low chemical emissions/ formaldehyde-free

ENGINEERED WOOD with LOCTITE® HB X PURBOND
FORMALDEHYDE-FREE | NO MIXING | NO WATER WASTE | NO SOLVENTS | NO CRACKS | NO CREEP

www.henkel-adhesives.com/engineeredwood

CPSIA information can be obtained
at www.ICGtesting.com
Printed in the USA
LVHW072336260322
714440LV00003B/3